PRAISE FOR THE BARBER, THE ASTRONAUT, AND THE GOLF BALL

T0283752

"A fascinating book, where the question [
than the answer and providing an intriguir

—Rob Parrish MD, PhD, neurosurgeon, board member, Lone Star Flight Museum

"*The Barber, The Astronaut, and The Golf Ball* is a story about history, fate, space and surprising connections. And it is beautifully rendered from all of those angles."

— Wil Haygood, author, The Butler

"An incredibly human, down-to-earth memoir of early space travel with a focus on friendship, family, and legacy. From Alan Shepard playing golf on the moon to both authors' fathers' behind-the-scenes contributions to astronaut technology, this book is a thrilling, eye-level recollection of humanity's greatest achievement."

— Keaton Patterson, lead buyer, Brazos Bookstore, Houston

"The personal side of Alan Shepard—the first US astronaut and fifth guy to walk on the moon—and his genuine friendship with Carlos the Barber. Newer and younger residents of Space City will connect with the great stories as we go forward on the next exploration of the moon."

—Jon Powell, scientist, city councilman, former mayor, Taylor Lake Village, TX.

THE BARBER, THE ASTRONAUT, AND THE GOLF BALL

ALSO BY BARBARA RADNOFSKY

Listening Space (Amazon, 2021)

A Citizen's Guide to Impeachment (Melville House, 2017)

Stepping Forward (Lulu, 2007)

The Dancer's Dead (Lulu, 2006)

THE BARBER, THE ASTRONAUT, AND THE GOLF BALL

BARBARA RADNOFSKY

ED SUPKIS

Stoney Creek Publishing

Published by

Stoney Creek Publishing Group

StoneyCreekPublishing.com

ISBN: 979-8-9891203-4-5
ISBN (ebook): 979-8-9891203-5-2
Library of Congress Control Number: 2024915651

Cover design by Market Your Industry, MarketYourIndustry.com. Cover image from "The Barber, The Astronaut, and the Golf Ball" directed by Jonathan Richards https://vimeo.com/878485586. Used by permission.

Photos page x and 72: Jonathan Richards. Used by permission.

Photos page 46, 136, and 143: Brian Snook. Used by permission

Photo page 107: Villagomez family photo

Photos page xxiii, 90, 152: NASA

To our fathers, Matthew Radnofsky and Daniel Supkis, lifelong friends, who worked together and conspired together at NASA's Manned Spacecraft Center Division of Crew Systems.

CONTENTS

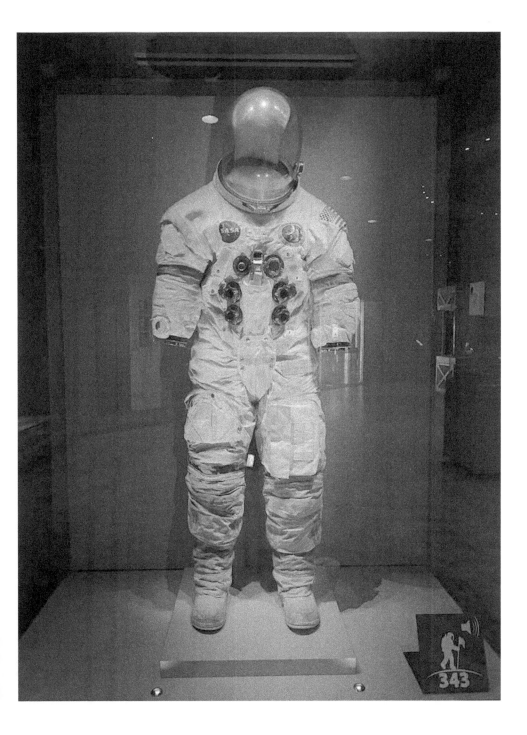

343

INTRODUCTION
A SEA STORY

This is a "sea story" lovingly told—for the past fifty-plus years—by our Navy combat veteran friend Carlos Villagomez, longtime barber to the astronauts of NASA and the owner of Carlos Barbershop and Beer Garden in Webster, Texas.

Our barber enjoyed a long friendship with the first American in space—astronaut Alan Shepard, the only one of the "Original Seven" Mercury astronauts to walk on the moon as part of the fabled Apollo program. At the end of the successful 1971 Apollo 14 moonwalking exploration, the already famous first American in space played golf on the moon to the surprise and delight of many people on the planet.

Afterwards, Shepard gave Carlos a memento of that voyage: a golf ball, autographed to Carlos.

THE ASTRONAUT, A "SENTIMENTALIST"

History books provide much information about U.S. Naval Academy graduate Alan Shepard, who served in a fighter squadron and several tours on aircraft carriers, and in a variety of positions—

including as Navy test pilot, operations officer, test pilot instructor, and aircraft readiness officer.

His daughter, Laura Shepard Churchley, captured her complex, intelligent, heroic father in a beautiful word: "sentimentalist."

In our childhoods of the 1950s and 1960s, the word was a loving tribute to a confident father capable of tender emotions, while performing his other duties in the greater world. In the Manned Spacecraft Center community where we, Barbara and Ed, grew up as children of NASA personnel, many with military backgrounds and with a devotion to the scientific method, the word implied no weaknesses or excesses.

In 1959, the new National Aeronautics and Space Administration chose seven men (the "Mercury Seven") from America's top 110 test pilots to serve as the country's first "astronauts." NASA introduced them to the world—with a barrage of publicity—as they began intensive training.

NASA selected Alan Shepard (1923-1998) as one of the Mercury Seven and as the first American to fly in space. The other six were Walter Schirra (1923–2007), Virgil "Gus" Grissom (1926–67), Scott Carpenter (1925–2013), John Glenn (1921–2016), Donald "Deke" Slayton (1924–93), and Gordon Cooper (1927–2004).

Shepard's 1961 suborbital flight—twenty-three days after the Soviet Union successfully launched the first man in space, cosmonaut Yuri Gagarin—brought him with his wife, Louise, to the world's attention with a massive parade and a famous meeting at the White House with President and Mrs. John F. Kennedy. A lifetime of press scrutiny would follow. Shepard and his family learned to manage the attention, balancing their public responsibilities to NASA and the American people as well as their private duties to family.

THE BARBER, A NAVY MAN

Carlos Villagomez enlisted in the Navy as a teenager in the aftermath of World War II. He served as a seaman, a bosun's mate aboard a large gunboat, in charge of the captain's boat for travel—the captain's "gig." Carlos saw combat, assisting frogmen getting to and from missions on the seas and rivers of Indochina, patrolling and rescuing thousands of fleeing civilians under fire in the north of what is now Vietnam, and transporting them to safer locations in the south. He also volunteered as an observer to the effects of an underwater atomic bomb detonation within two miles of a ship and her crew. Carlos explains that it was an awesome experience aboard, with a spectacular waterspout. They were evacuated as their old test ship began sinking.

In the early 1960s, NASA recruited Carlos to leave his barbering work in Houston and set up shop to become part of a new community—a new space city—temporarily organized in South Houston, as the government built a modern spacecraft center on a cow pasture farther south, toward Galveston.

Carlos has lived a life of service, as a brother and son in a remarkable family raising 15 children, in naval combat serving his country, in his profession as a barber, and in his community as an elected official, an active church member, and a generous and innovative Rotarian, businessman, father, grandfather, and great-grandfather.

We (your authors, married since May 1982) regard Carlos as much more than a barber. He remains our closest and oldest friend, in effect part of our family. He's so close to Ed that it was Carlos's sad responsibility to inform Ed of the deaths of Ed's parents.

Carlos may not be famous outside his community, but he's beloved and respected in the town of Webster and surrounding parts.

THE GOLF BALL

The golf ball is just that—a golf ball. It bears the autograph of Alan Shepard, who often autographed golf balls, especially on golf courses. Respected space memorabilia expert and historian Robert Pearlman authenticated the signature. The inscription on the ball reads:

"To Carlos Alan Shepard"

A half-century after Shepard gave Carlos the ball, Carlos gave it—a bit gray with age, with the inscription turned pink—to his dear friend, Ed.

At the time Shepard commanded the 1971 Apollo 14 lunar mission, your authors were members of typical NASA families; we were both fourteen years old.

DAN SUPKIS

Ed's dad, Daniel Edward Supkis—a gifted athlete six feet and eight inches tall—grew up in the 1940s and 1950s playing street basketball in New York City with future greats like Bob Cousy. Dan's immense talent and work ethic were obvious. He knew hunger in the Depression, which only increased his ambition and drive.

Dan Supkis came from a long line of brilliant scientists and engineers, and the tall, thin, intense young man earned a full scholarship to Georgetown University. Dan Supkis was Georgetown's basketball team captain in 1951–52.

After he graduated with a degree in chemistry, the basketball pros came calling. Recruited by the Boston Celtics, Dan chose instead to become an engineer, as did his older brother Stanley. While working for Norton Abrasives Company, Stanley invented and improved their line of sandpapers, earning many patents.

Their dad—Stanley Sr.—was a talented machinist for the

Sperry Corporation, making Sperry gyroscopes during World War II. Sperry engineers routinely consulted the extraordinary machinist on engineering and design issues.

In the late 1950s, Ed's dad moved his family from Marshall, Texas, to Texas City, to become the production manager for the Monsanto Styrene Unit, within recent memories of the 1947 deadly Texas City explosion. After a few years in the heavily polluted area, Dan, Jane, his strikingly beautiful and intelligent wife, and their two sons moved twenty miles north. They chose a new housing development called El Lago, across Taylor Lake from another small community called Timber Cove. Both communities were close to the cow pasture that would be transformed into the Manned Spacecraft Center. The two subdivisions would house many of the early NASA folks, including astronauts.

When Dan heard of a new spacecraft center coming to Houston, he signed up immediately and began working with Barbara's dad at NASA's Crew Systems Division. The collaboration brought about extraordinary results.

On January 27, 1967, three astronauts—Ed White, Roger Chaffee, and Gus Grissom—died in a fire that broke out in the capsule of Apollo 1 as it sat on the launchpad. The tragedy shocked our families just as it shocked the space community and the country. Dan Supkis came home and raided his son's chemistry set, using the chemicals to begin developing new fireproof coatings and materials.

He brainstormed in the family garage, testing the materials with a Bernzomatic propane torch. By the end of the weekend, Dan had samples to show Matt Radnofsky. They developed a remarkable fluoroelastomer called Fluorel.

In addition to being fireproof, Fluorel was really tough stuff. NASA used Fluorel for many purposes, including moon boots and astronaut gloves. Our dads' work was put to the test during the Apollo 13 emergency—when an oxygen tank ruptured, disabling electrical and life-support systems and putting crewmembers' lives in jeopardy. Their coatings of fluoroelastomer safely protected the electrical systems once the crew restarted them. During the long return to earth, these coatings prevented the key electrical systems from shorting out.

After the Apollo 1 fire, NASA coated every possible electrical component with Fluorel, particularly the backs of circuit breakers and switches. During the Apollo 13 command module re-entry into Earth's atmosphere, there was considerable concern that ice, which had accumulated in the command module, would melt and "short the capsule out." Fluorel prevented any water from entering any electrical components; Apollo 13 experienced no such shorts or fires.

Fluorel coatings enabled a safe re-entry and splashdown. And descendants of Fluorel are in use today—Apple watch bands are made from a fluoroelastomer.

MATT RADNOFSKY

Matt Radnofsky, Chief of Crew Systems, described his work to his kids, as we were growing up, as attending to everything the astronauts could touch and everything from the skin of the spacecraft inward—from the freeze-dried ice cream and Tang Orange drink powder to the undergarments and space suits and life rafts to the switches and wires. We knew his life was dedicated to flight safety.

During the Apollo 13 crisis, Matt was locked in a conference room with the folks working with the same materials that the Apollo 13 astronauts had at hand. Matt was tasked with addressing a problem with the carbon dioxide scrubbers. The astronauts had to remove carbon dioxide from the air they were breathing or they would die of CO_2 toxicity before they got home. But the lunar module's scrubbers were almost used up.

Matt and others working on the problem came up with a contraption, using a logbook cover and duct tape—to jury-rig the scrubbers and bring the crew home safely. The Tom Hanks Apollo 13 movie portrays this situation very well.

Matt Radnofsky was great friends with Dan Supkis. The two friends became our matchmakers and, eventually, our fathers-in-law.

Our dads—plus the thousands of other dedicated, safety-conscious, innovative folks working on space suits, gear, coatings, and myriad other key aspects of the Manned Spacecraft program—accomplished a clear mission set forth by President Kennedy: to put a man on the moon before the end of the decade of the 1960s. They focused on safety as they also prioritized the astronauts' safe return to the earth.

Matt had a lifelong devotion to flight safety, based on his experiences during World War II. From 1942 to 1947, he served in the Army Air Corps—which later became the Air Force. (You can read more about Matt's experiences in the materials cited in the Sources at the end of this book.)

On November 21, 1944, Lieutenant Radnofsky was flying his twenty-fourth mission over Germany, serving as a lead navigator for an element, or group of planes, that was hit repeatedly by flak at their target. The pilot of the B-17 ordered his crew to bail out, but Matt was partially paralyzed. Jagged metal flak cut through his flak jacket and lodged in his back. The impact also caused his parachute to spring open.

The plane's bombardier, First Lieutenant Douglas K. McKnight, removed his own parachute and put it on Matt, who was unconscious and bleeding on the floor. The bombardier slapped Matt awake and muscled him out of the damaged plane before finding another chute for himself and following him.

As he descended, chute deployed, Matt was hit by enemy machine gun fire. The bullets would remain in his back, near his spine, for the rest of his life.

In 1992, Matt recounted what happened next in a report for the parachute manufacturer:

... I was grateful for the "new type" of single point harness's "Twist and Hit" release, as it was the only way I could extricate myself from entanglement on landing, being partially paralysed [sic] when being machine gunned. I was then captured by the German populace and sent to a hospital in Veckta, Germany.

I was subsequently released to a German P.O.W. Camp, Stalag 11-B, three months later, transferred to Dulag Luft Wet.31 AR from which I left, reaching Allied lines near Frankfurt. I was flown to England on repatriation on the 12th of April 1945.

M.I.R./15 June 1992

Matt emphasized that the parachute was the only piece of safety equipment that worked properly. He had just turned twenty. He committed his life's work to aviation safety.

He escaped POW camp with a small group of other prisoners as the Germans were in disarray toward the end of the war. The young men stole German uniforms and a wood-burning Jeep. (Matt told his children of Germany's crippling fuel shortages and conversion of vehicle engines to the use of readily available fuel in the form of wood.) Matt's knowledge of German aided the POWs in reaching Allied lines near Frankfurt, where they traded the Jeep for a plane ride to London in April 1945. In London, they were sent to jail because they were wearing partial German uniforms.

Matt's commanding officer came to London and identified him, and brought him back to a base in Thurleigh. Matt recommended McKnight, the bombardier, for the highest commendation—the Medal of Honor—for combat heroism. (The Air Force lost the letter, which remained buried for twenty years. The two men—bombardier and navigator—met again in the 1960s at Ellington Air Force Base, and McKnight—by then an Air Force colonel—received a belated Silver Star for his extraordinary valor in saving the young lieutenant.)

Matt was then sent stateside to a hospital, which he promptly left to find and marry his childhood sweetheart, Eunice. When they presented themselves at city hall, their wedding plans were thwarted because Matt was too young to get married without written parental permission, which came swiftly.

In 1948 and 1949, thanks to the U.S. government and the G.I. Bill, Matt earned his bachelor's and master's degrees from Boston University. He often praised the G.I. Bill for changing his life, providing generous benefits in proportion to combat injuries, allowing him a first-class education.

From 1950 to 1955, Matt worked at both the Naval Aeromedical Equipment Laboratory at the Philadelphia Naval Yard, and as a designer and tailor for safety gear and sportswear for L.W. Foster Sportswear. Foster also was developing safety gear for the Department of Defense.

Matt's constant focus was aviation safety. His files reveal information on inventions and designs for pressure suits, life rafts, vests, helmets, gloves, lighting systems, radar balloons, reflection units, eye protectors, and oxygen masks—and there are photos of Matt in icy water, demonstrating the designs. He worked with the U.S. Navy to design rafts and life vests for downed aviators.

Matt moved his family to Newport News, Virginia, near NASA's Langley Research Center. He designed, constructed, sewed, tested, and developed equipment, suits, helmets, materials, coatings, and clothing—everything needed for humans to explore and survive space and safely return home to Earth. Matt never stopped planning for space travel. (His papers, patents, writings, and drawings from work when he retired show he was also deeply engaged with top NASA scientists and astronauts in efforts to plan and execute safe manned travel to and exploration of Mars.)

When NASA's Space Task Force transferred from Langley to Texas in the early 1960s, Matt joined NASA, moving the family to South Houston, where NASA rented temporary offices and research spaces, as personnel awaited construction of the Manned Spacecraft Center even further south.

The Radnofsky family arrived in Texas in 1961 to a house filled with mud and snakes in the wake of Hurricane Carla. The new home created much excitement for four little kids looking for adventure, and a much different world for their beautiful, stalwart, and intelligent mother.

NASA then moved Matt from temporary office buildings in South Houston to the current location of the Manned Spacecraft Center area, with buildings still under construction on a massive cow pasture in a rural area between Houston and Galveston, near Webster, Texas. During his career at NASA, Matt served as Head of Survival Equipment, Chief of Apollo Support, and Chief of Crew Systems.

The Johnson Space Center drew adventurous, brilliant, dedicated scientists, mathematicians, and support personnel, who moved to a rural location to build a new Space City.

NASA folks like our parents and their colleagues remain our heroes. Barbara admits to romanticizing these people since earliest childhood.

Ed worked at NASA as a summer intern. Intent on being a doctor, he helped review the content of medical kits used in space. Barbara's first summer job as a teenager—in the early 1970s—was behind the cash register at the souvenir counter in the NASA cafeteria.

NASA also recruited barber Carlos Villagomez to become a part of the new community. Through many decades and countless haircuts—he's now cutting hair for a fourth generation of our families—your authors have loved listening to Carlos's barbershop stories, particularly tales of the people who worked in the manned space program.

We also admit to bias in approaching this story.

We love the subjects, including Carlos, who is very much a part of the greater NASA family. Barbara has also served as occasional test dummy (her hair grows back fast) for determining suitability of fledgling associate barbers at the Carlos Beer Garden and Barbershop.

We knew from an early age—from stories, memorabilia on the barbershop walls, and gifts from Carlos—that Carlos was good friends with Alan Shepard.

We certainly enjoyed NASA memorabilia as kids and as adults. Long ago, Carlos gave Ed a Shepard photo—autographed to Ed by Alan Shepard with a felt pen of the day—which sadly faded but still proudly adorns our wall.

Matt's NASA job included responsibility for material selection, testing, and designing astronauts' personal equipment and safety equipment, including life rafts (tested in our community swimming pool with the aid of happy children), and flotation collars for the capsules—anything the crew might need. Of greatest interest to the public were the space suits.

And, as in every NASA family, we had interesting NASA stuff that was too beat-up for proper use. Barbara's dad brought home an old greenish flight suit, designed to be worn by crews for comfort, leisure, or messy work—one of the beat-up ones, not intended for public appearances but for safety and comfort. Safety, including protection from fire, was paramount to our dads. The beat-up suit still bore the astronaut's name tag.

"Flight suits" came in various materials and styles over the years. The most famous and familiar NASA flight suits are shades of blue, with handsome patches; the early suits were manufactured by Lou Foster, the company for which Matt worked after the war designing safety gear and sports clothing. The working flight suits are far more drab.

If you ever get to see the earliest astronauts' photogenic flight suits, check out those fancy coveralls' labels. Those first NASA blue flight suits bear the label: "Fosterwear styled by Lou Foster." In the Sources at the back of this book, you'll find the story of what became of one of John Glenn's Friendship 7 flight suits, gifted to

Bill Green, a friend of Matt's who now works at the Evergreen Aviation and Space Museum in McMinnville, Oregon.

As with many of Matt Radnofsky's designs, these coveralls, with lots of pockets, drew much inspiration from military coveralls he'd worn in England and Germany during World War II. He learned much about flight safety needs from his military experiences in air combat while sitting on a piece of sheet metal—as an improvised flak guard (which protected him no better than his useless flak jacket).

In short, our dads used their life experiences to innovate. They received superb educations in science—one working in chemistry and confronting the dangers of chemical plants, the other studying aviation disasters to improve safety. Matt Radnofsky's designs and materials and Dan Supkis' chemistry expertise became critical to astronaut survival, as Scott Carpenter explained in his memoir, *For Spacious Skies*, and in a note to Matt.

CARLOS THE BARBER AND ASTRONAUT MEMORABILIA

Carlos the barber collected astronaut memorabilia from the beginning of his career as barber to NASA personnel, when his customers began a tradition of bringing him souvenirs and flown objects. Carlos's barbershop displays an unending, rotating series of gifts from his astronaut friends—including featured items signed by Alan Shepard.

Carlos and his trusted associate barbers update the walls as returning astronauts keep them current, but the multiple Shepard-signed photos and posters have been a constant presence.

We NASA kids never forgot that in 1971, Alan Shepard golfed on the moon at the end of his Apollo 14 moonwalk. It became one of the most famous sporting moments in world history.

We kids heard stories that Shepard had signed various golf balls over the years. Ed best knew Carlos's story about a particular golf ball, a gift Carlos explains he received after Shepard emerged from his post-mission quarantine in 1971.

Shepard's brief golf episode eclipsed the famous Annapolis graduate's career as a naval officer, pioneer in carrier takeoffs, fighter, and test pilot, and original Mercury Seven astronaut. Enshrined in history as the first American in space, Shepard was also known in his NASA community as a successful businessman, husband, father of what he proudly called three "gung-ho" girls, and grandfather.

We knew a few basics: Alan Shepard left two golf balls on the surface of the moon after his golfing display. We knew that NASA engineer Jack Kinzler, who would later design Skylab's sunshade, crafted a golf club head that would attach to Shepard's tool for gathering moon soil samples. Shepard could quickly and easily convert the tool into his famous moon golf club.

At the end of his successful mission, Shepard had needed an extra golf ball—and produced a second from one of his pouches—having whiffed his first attempt at a swing.

He had practiced his swing on earth, but not under moon conditions. He clearly knew he would face immense difficulties in the attempt. After he missed the first shot, he quickly produced a "reserve" ball, which he dropped and successfully hit.

Did Shepard have yet another backup, a third ball waiting in reserve? If so, did he later give the unused ball to his barber? The background of the "third" golf ball and its story was a mystery we wanted to explore.

For the folks who love Carlos, the story of the Carlos golf ball isn't really about a golf ball at all. It's a tale of the wonderful friendship between two Navy men—one a bosun's mate and the other an admiral.

It's a weathered sea story—a tale held together by the truth of two men's deep friendship and the fond memories of a barber who cut the hair of a man who walked and golfed on the moon.

CHAPTER 1
THE ASTRONAUT

THE NIGHT before Alan Shepard entered Apollo 14 quarantine, the astronaut and his barber, Carlos Villagomez, sat outside Shepard's home, relaxing by his pool under the night sky. They talked over drinks. Very late that evening, Shepard would be picked up to be isolated from possible earth bugs at the start of his mission to the moon in early 1971.

Carlos had finished cutting Shepard's hair extremely short in the astronaut's living room, as Shepard's wife, Louise, watched with great interest—it was the first time she'd seen her husband having his hair cut.

The two men relaxed, had a few drinks, and looked up at the stars and the moon. Shepard directed the barber to ask him some questions; Carlos understood his friend wanted to practice privately for press and dignitaries. Carlos was happy to oblige. He asked: "What will you do up there?"

Shepard—eight years older than Carlos—guided his friend through the night sky as he described the landing spot on the moon and ran through the plans, omitting any reference to the one activity that was a closely kept secret: a golf demonstration, contingent on all other tasks being successfully completed on time. Not even Shepard's family knew of the plan.

The astronaut told Carlos to look at his face as if it were the moon above them and then pointed to the spot on his face where the landing would occur. Late into the night, sharing drinks under the stars, Carlos asked questions.

Shepard's fortitude amazed Carlos, who knew Shepard had been grounded by an ear ailment after his suborbital flight a decade earlier. Shepard had persevered, found a cure, and endured the extraordinarily tough training to master complex, grueling tasks—physically and mentally—for piloting the landing craft and conducting geological explorations. He had conquered much in preparing to command and carry out this mission.

Carlos marveled at his friend's confidence, quiet courage—and happiness. Shepard seemed calm and ready.

"He worked hard to try to get back on and now he's back on it," Carlos recalled. "And so I cut his hair, and he was saying, 'Ask me questions. I just want to exercise my ear,' or something like that. So I would ask him all kinds of dumb questions like, 'where will you land?' And I would say, 'Are you a little afraid?'"

But Shepard seemed serene, with no fear of the risks he would soon take. He told Carlos that he reached a point where "you're just ready to go."

Carlos asked Shepard to do something for him, like write his name in the lunar dust. Shepard didn't reply. He just gave his friend a wide grin.

WHO WAS ALAN SHEPARD?

We gained much insight from Shepard's 1994 book, written jointly with his friend and fellow Original Seven astronaut Deke Slayton; his 1981 interview;* and our series of interviews with Shepard's

---

* Shepard, interview, 1981 (https://m.youtube.com/watch?v=G3tFb543lsg)

daughter, Laura Shepard Churchley—who has also flown in space —as well as experts, NASA, and military veteran pilots who either knew Alan Shepard or knew about him as boss, colleague, and friend. Additional sources (including many firsthand, detailed accounts) are listed at the back of the book. They provide a portrait of an admirable man.

Churchley remembers her father as loving and sometimes stern. He was a "sentimentalist" who adored and took pride in his three daughters, she said. Shepard had been a talented and highly skilled Navy combat and test pilot. Around NASA, he was quiet, hardworking and intelligent, and in later years, when he ran the astronaut corps, he was considered a level-headed boss. He could also be a generous benefactor to folks in need, and although he was a deeply private man, he also understood the responsibilities that came with his public persona, which extended to his physical appearance.

And he loved golf.

SHEPARD'S DISCIPLINED CHILDHOOD IN 1920S RURAL AMERICA

Born in 1923, Alan Shepard was a rural New Hampshire boy whose father, a retired Army colonel, instilled in his son what Shepard described as a love of tinkering, building, and creating things that flew, rolled, or just plain hummed along—particularly model airplanes.

The remarkably intelligent boy received all his primary school education in a one-room schoolhouse that combined all six grades of farm community children. One extraordinary teacher handled all teaching duties—with strict discipline. The setting allowed Shepard to eavesdrop on the lessons of the grades above him, and he completed all six grades in five years.

Shepard credited his self-discipline and success to his upbringing in a family of hard workers in a rural community, with a sense of family and family achievement, and—with great emphasis —to "that one lady teacher" responsible for his childhood education when he was drawn to math and the physical sciences.

He analyzed his primary teacher's outsized influence in his life, describing it in a 1981 Academy of Achievement interview, after he had become a historical figure, celebrated moonwalker, and Navy admiral. He reflected on the source of his self-discipline:

A key thing was ... that one lady teacher in grade school, all sixth grades, one room, twenty-five students. She was about nine feet tall, as I recall. And a very tough disciplinarian, always had the ruler ready to whack the knuckles if some-body got out of hand and she just really ran a very well-disciplined group .... And it's interesting, I think most of the youngsters responded to that. There were one or two that couldn't handle it, and obviously they dropped by the wayside. But that still sticks out in my mind. That lady taught me how to study and really provided that kind of discipline, which is essentially still with me—I think certainly some of the characteristics which were helpful to me in the aviation business and the astronaut business were developed in those days.

Growing up in the 1920s and early 1930s, young Alan would rush to the local airport for after-school jobs and the chance to be near airplanes. He wanted to learn to fly.

Shepard's intelligence, upbringing, and studies enabled him to pass the entrance test for the Naval Academy at Annapolis, from which he graduated and received his commission as an ensign on June 7, 1944—an early commission because of World War II. He was attached to a destroyer, the USS Cogswell.

He was a Navy flight instructor assigned to a fighter squadron in the Mediterranean and received test pilot training at Patuxent River, Maryland. He took part in a wide variety of aviation testing and experiments and pioneered in-flight refueling and using Navy ships for aviation. He valued his training, skill and calm in carrier landings.

"There are few things that shake up a land pilot more than the prospect of searching an angry ocean for a bobbing and rolling gray slab that is an aircraft carrier deck," he says in *Moon Shot*, the memoir he wrote with fellow Mercury astronaut Deke Slayton.

Despite strong Navy ties, including lifelong friendships, Shepard changed his career plans—a choice clearly regarded by his father as a judgment error—and applied for an entirely new kind of service to the country—as an "astronaut." (Later, his father admitted he'd been wrong.)

In April 1959, Shepard became one of the test pilots chosen by NASA for the Mercury program. Two years later, President John F. Kennedy committed the country to land a man on the moon in the 1960s. Shepard would play important roles in achieving that goal, and profound, ongoing progress in education and research in science and mathematics.

Laura Shepard Churchley offered some insights into what it was like growing up with her famous dad.

> He was very fatherly, I'll tell you that much. We did have to be obedient. That was expected of us. He had a fun sense of humor ... and at the same time could be very strict .... I do remember if we misbehaved and Daddy wasn't home at the time, when he did come home, we would get a little spanking. And Mother wouldn't do it. She would wait for Daddy to come home. And we learned very quickly that he was serious when he meant we couldn't do something. We didn't do it. We learned. We learned the hard way.

Laura recalled that, after he left NASA and started his own company, her father took care of his employees.

The wife of one of the employees was in the hospital and Daddy helped with the financials of whatever it was that she needed. He was there. He was helping his employee's wife. Those are the kinds of things. And he didn't announce it to anybody. He never said, "Oh, I'm doing this." When he saw a need, he took care of it. He was very generous.

Laura said that he relaxed more after retiring from NASA, enjoying life out of the constant limelight of the space program. "After he retired, he didn't like all the publicity," she said. "We were a pretty private family. We were kind of in the background."

BALANCING FAME AND PRIVACY

Alan's wife, Louise Brewer Shepard, also valued and guarded the family privacy. She and Alan were aided—from the beginning—by NASA press man Gene Horton, who balanced privacy concerns with the world press demanding to interview the first American in space. As with all astronaut wives, Louise had to defend her family from the press.

In their joint memoir, *Moon Shot*, Shepard and fellow astronaut Deke Slayton note Louise Shepard's effectiveness in protecting the privacy of her family. During the lead-up to her husband's historic sub-orbital flight, she posted this notice as a clear and bold sign to the world's press surrounding her home, and not permitted inside:

THERE ARE NO REPORTERS INSIDE. I WILL HAVE A STATEMENT FOR THE PRESS AFTER THE FLIGHT.

The family heard the sounds of retreat each time the reporters approached the house and saw the sign.

After her husband's safe landing, Louise met him on the East Coast for "all the parades and whoopla to follow the next morning," as NASA press man Gene Horton wrote in his memoir *Losing Them*.

The whoopla included a meeting with President John F. Kennedy and his wife, Jackie, in a much-photographed Rose Garden ceremony.

Alan Shepard, calmly awaiting recovery in the Atlantic Ocean, had returned from space to extraordinary press attention. Horton—who began his career as a newspaperman—knew that the quick trip and safe splashdown of Shepard's Freedom 7 capsule meant his work was just beginning.

The young NASA press man travelled from Florida's Cocoa Beach "in a small single-engine plane which was packed to its sides with reporters—a small media pool" to attend Shepard's history-making trip.

Horton and the reporters began "a long recovery watch, stringing communications and camping nights beneath the stars in humid bedrolls, or if more fortunate, inside the single beachfront Quonset hut."*

Horton listened as the launch was announced by NASA spokesman Shorty Powers, who coined a new expression—"AOK" —in advising the world that Shepard's Redstone rocket was "clear of the beach and the clock is running." Minutes later, Horton noted, "Freedom Seven was safely back at sea and bobbing about in the Atlantic," Horton wrote. He and the reporters piled into another plane and reached the mainland to file their stories.

Horton fielded "frenzied questions of the press ... from the informed to the ridiculous ('Did he do as good as the chimp?')"—a reference to earlier test flights that used monkeys.

Horton was committed to protecting the privacy of Shepard and his family. "By late afternoon the same day, I was aboard a

---

* Horton, *Losing Them*, 194–97

second flight to Langley Field," he recalled later. "Shepard had earned time away from the cameras to enjoy, however briefly, with his wife—far away from the prying eyes of the press.

"It was my job to help make this happen. Once we landed, he met his comely wife Louise, and I took them to a secret hideaway to enjoy quiet time together before all the parades and ... scheduled meeting with the president."*

"ONLY ONE MAN CAN BE THE FIRST IN SPACE"

Alan Shepard Met Challenges Aided by Fellow Mercury Astronaut Deke Slayton

In mid-1960, a Redstone rocket carried an unmanned Mercury capsule on a perfect test flight, and NASA knew the program was ready for a manned flight. Robert Gilruth, chief of the Space Task Force Group, who had the final say on who would fly in space, gathered the Mercury Seven.

"Everybody better start thinking about who goes first atop a Redstone rocket," he said, according to *Moon Shot*. "I want you guys to take a peer vote. If you couldn't make the first flight, select the man you think should go."

Gilruth told the astronauts to write their choice on a piece of paper and leave it in his office. On January 19, 1961 (the day before Kennedy's inauguration), Gilruth again summoned the Mercury Seven to the astronaut office. He praised them for their work but added, "you all know only one man can be first in space."

After a long pause, he announced: "Alan Shepard will make the first suborbital Redstone flight."

"Deke Slayton may have been the only one among the six not selected to be pleased it was Shepard," the two friends wrote in *Moon Shot*. "Slayton's private list had himself pegged as first choice, and Shepard second."

---

*   Horton, *Losing Them*, 197

Slayton was the only one of the original seven astronauts who didn't fly in the space program's first decade. He was grounded because of a minor heart irregularity, which only strengthened his friendship with Shepard, who was also grounded after his first flight because of Méniére's disease, an inner-ear issue that causes dizziness and tinnitus. Laura Shepard Churchley noted both men battled their ailments to regain flight status.

> Daddy was grounded after his first flight for about six years .... He was upset that he was not able to fly. That was a very hard time in my mother's life, having him home and [him] not being very happy. So we all were very relieved when his ear was fixed properly because he could get back to normal.
>
> When he was chosen—when he got his ear fixed, that allowed him to be in contention. To be selected to go on a flight.

A MERCURY FRIENDSHIP

On his short *Freedom 7* flight, Shepard wrote in *Moon Shot*, he drew strength from Slayton's voice. As they worked closely together in the Mercury days, a deep friendship developed between the two military test pilots. A decade later in the Apollo era, after Shepard had surgery to fix his ear and returned to flight status, Slayton's heart issue, atrial fibrillation, kept him grounded.

Slayton wrote in *Moon Shot* that Alan was going to the moon "for all seven of them." He also revealed his concerns for his friend as he "walked the tower"—making a test pilot's instinctive preflight plane check. For astronauts, the "walk" differed. Slayton visited the massive Saturn rocket that would launch America's first man into space to the moon. He took the elevator four hundred feet up to launchpad 39's "penthouse" to "kick the tires" himself before the Apollo 14 flight.

Slayton acknowledged knots in his stomach for "the special space reserved for best friend." Shepard was sleeping soundly, yet Slayton admitted to fear. He "didn't know if he had it in him to bear the loss of Alan Shepard." Shepard recalls kissing Louise goodbye through protective glass panes of quarantine and saying with a smile, "I won't be making my usual phone call tomorrow night. I'll be leaving town."

So close was the friendship between the two men that Shepard sacrificed his legendary desire for privacy and agreed to a joint memoir. Deke Slayton's widow Bobbie made clear in a biography of Shepard (*Light This Candle*, by Neal Thompson) that Shepard had agreed to co-write the memoir *Moon Shot* to assist Slayton financially. By then, Slayton had been diagnosed with brain cancer and needed the money.

"He did that for Deke and me," she said.

"WHERE ALL MY FRIENDS WERE ..."

In Mission Control, as Slayton watched his friend on the moon, lifting his arm and describing the lunar dust, he noticed the manner in which Shepard had turned on the moon's surface and stopped. While observing this momentary action, Slayton wrote, suddenly he felt that a "powerful emotional strain was running through this man." Slayton then noticed that his friend was looking at Earth.

Shepard later explained his feelings: As he stood on the moon's surface, he stopped for a long moment. The Earth was home, "where all my friends were .... "

Shepard wrote in *Moon Shot* that he realized tears streaked his cheeks. He and the entire team had encountered serious complications—and solved each emergency. He calmly observed that any tension resolved and faded "before that utter, fragile beauty of our planet," leading him to conclude "everyone must learn to live on the planet together."

In *We Seven*, the 1962 autobiographical accounts by the original Mercury astronauts, the book's editors spend several pages following Shepard's account of his *Freedom 7* flight explaining how and why NASA's thorough examining physicians showed "unconcealed admiration" for Shepard.

After the historic flight, the examining psychiatrists remarked on the "subject's" calm, self-possessed attitude, and his cheerfulness—indeed delight—that his performance had been better than the astronaut had expected of himself.

The doctors noted Shepard "was more concerned about performing efficiently than about external dangers," and he had "consciously controlled" what the astronaut self-reported as moderate apprehension during the preflight period. Shepard explained his ability to control his concerns "by focusing his thoughts on technical details of his job," resulting in "very little anxiety" during the immediate pre-launch. Moreover, the doctors noted, the astronaut's only feeling of concern was rational—when he fell behind on one of his tasks, during manual control at re-entry.

Other physicians thoroughly analyzed all other data they could obtain from the sensors attached to his body. They made a similarly positive report, determining that Shepard had sweated off three pounds of his body weight from perspiration in his hot, rubberized spacesuit. They found him to be "a patient and understanding guinea pig."

Throughout the massively detailed exams and tests, there's only one report of a comment from the test subject—concerning the taking of countless body fluid samples before and after the flight. "I felt," he said with a smile to the doctors, "as though an unusual number of needles were used."

The Apollo 14 moon landing had been fraught with problems, involving a series of newly required actions to start the engine at low power, to enable guidance and steering, to disable an abort program, and lock into landing and descent software. The astronauts and Mission Control were relying on an emergency patch on which the astronauts had no time to practice.

NASA Flight Director Gene Kranz wrote in his book *Failure Is Not an Option* of Shepard's great calm and strength as the moon landing proceeded: "Shepard sounded just as he had all those years ago when he first went into space. Marvelously calm, his voice was flat and emotionless as [Edgar] Mitchell read the checklist, verified the switches, and entered data into the computer."

Then, with one major problem solved, more complications arose, as the radar data didn't come through. After a stressful period, the astronauts cycled the circuit breaker and obtained radar data just in time. Afterwards, Shepard said, "I had come too far to abandon the moon. I would have continued the approach even without the radar." Experts claim Shepard would have run out of fuel before landing had he flown without radar.

"But," Kranz concluded, "everyone who knew Al never doubted that he would have given it a shot. We also never doubted that he would have had to abort. The fuel budget was just too tight." Kranz ended his Apollo 14 chapter with great compliments to Shepard, including praise for the astronaut's "lyrical comment" as the calm crew bypassed the problematic abort program. Shepard stated: "It's a beautiful day to land at Fra Mauro."

Shepard was referencing the planned landing spot he knew well, which he'd described to his barber as so very familiar to him, thanks to tremendous practice sessions with extraordinarily detailed models. Alan Shepard was supremely prepared; he knew the land he approached.

"DON'T GET BACK ON YOUR HEELS"

Carlos repeatedly voiced his amazement at Shepard's calm readiness and quiet confidence. The night before quarantine began, Shepard emphasized to Carlos that his fearlessness was based on complete preparation, with an immense amount of time spent in quality training and practice.

Shepard was to repeat these concepts publicly. Ten years later, in his 1981 Academy of Achievement interview, he emphasized his astronaut work involved basically "research and development," and a commitment to time taken to practice, plan, and design for things that might go wrong, while maintaining objectivity.

This preparation, he explained, allowed for significant geological discoveries made in Apollo 14, overcoming various obstacles. Shepard also spoke of his childhood, schooling, and family background influences, instilling in him the importance of diligence, training, discipline, hard work, and staying on track, enabling him to achieve goals in his work in aviation and being an astronaut.

"If you want to achieve something," he said, "don't get back on your heels."

Carlos's observations from Shepard's pre-mission explanations track the astronaut's statements in an Apollo 14 NASA Technical Debrief cited by engineer-historian David Mindell in his book, *Digital Apollo*. The book details and analyzes the role of a unique class of pilots—Apollo lunar landing pilots—as those astronauts interacted with automated systems in six lunar landings.

Shepard consistently told his interviewers after his return from the moon that he'd had no problem finding the landing spot, which looked just like the plaster-of-Paris training model. What was the determining factor for Shepard in that success? The "high fidelity of the simulator display" and the time he spent in training.

Shepard and Slayton note in *Moon Shot* that Shepard had landed just feet from the "X" they'd marked on their landing chart during training. Their book also tells the story of what happened at the Shepard home at the time of touchdown on the moon. Louise shrieked with joy. And between laughing and crying, she told the family: "We can't call him Old Man Moses anymore. He's reached his Promised Land!"

When Shepard safely landed his craft and stepped onto the moon's surface, his words seemed to summarize the extraordinary teamwork, including quiet but dramatic efforts required to solve the host of last-minute technical issues and his years-long determination to overcome his own medical issues.

"And it's been a long way. But we're here," he said.

He spoke openly in the 1981 Academy interview of his time on the moon—as well as the great enjoyment he derived from it. He provided a natural and engaging grin during the interview. He spoke of achieving all goals set, in a timely manner, providing an educational demonstration for the worldwide audience watching on TV, with an experiment involving a sport he loved better than any other.

"I HAVE A LITTLE WHITE PELLET ..."

Alan Shepard loved golf, and he played frequently with celebrity friends, like the comedian Bob Hope, who was known for always carrying a golf club. Hope used his golf club as a cane, and a visit to Johnson Space Center gave Shepard an idea, according to Robert Pearlman, the space historian.

Shepard thought, "maybe I could take a golf club to the moon and maybe I could use it as a science demonstration to show the differences in one sixth gravity versus on Earth," historian and memorabilia expert Pearlman told us. "And he pitched it to his bosses, and they said no. They said, 'first, you can't take a golf club to the moon because it's too big, too heavy. You can't, we won't be able to accommodate a full golf club. So that's out of the question.'"

Shepard proposed modifying the contingency sample collector with the head of a six iron, arguing that it "all folds up nice and tight. And so, your math problem and your volume problem is solved," Pearlman said. "And then they said, 'Well, no. Okay, that's solved, but we're not going to let you go do it because we're facing an issue. The American public had lost interest in space exploration. We've just gotten through the Apollo 13 accident, which brought everyone new eyes on the program.' They [were] scared of what Congress would think of this multimillion-dollar mission doing something as frivolous as a string of golf on the moon."

Undaunted, Shepard proposed a compromise, as Pearlman explained:

> So Shepard said: "Well, okay, here's what I'll do. If anything goes wrong on the mission, I won't do it. And I'll save it till the very end of the moon walk so that we'll know that everything was successful. And then I'll do it on my dime. No taxpayer money spent. You know, you know, creating these tools or getting the golf balls. And I'll do it on my own."
>
> And finally, they relented and said, "you can fly the golf club and the ball." So, it wasn't done in secret. People have suggested over the years that he smuggled them on board, but no, they knew that they were flying.
>
> But they did, he did, keep it secret from the rest of the world until he was standing on the moon at the end of his last moonwalk.

In his 1981 interview, Shepard explained how the plan came together:

> And at the end of the of the second mission, I played with a little device that I had received clearance to play with before, and that was a makeshift golf club, the couple of golf balls up there. All of us wanted to try to think of something which would interest perhaps young people primarily in the difference in the gravity.
>
> The gravity is only one sixth—the lack of atmosphere. There's absolutely a total vacuum up there. And some of the other guys, the clever guys before us, had dropped the little lead ball and the feather and watched them slowly

proceed at exactly the same rate to the surface. And so some imagination had been used before.

And I'm—being a golfer, I thought, well, now, if I could just get a club up there and get it going through the ball at the same speed, that it was going to go six times as far there as it should have gone on the Earth.

So ... we designed the club ... to fit on the handle we had up there to scoop up dust samples with.

And I cleared it with the powers that be. We practiced in the spacesuit before we went to be sure there were no safety implications. And the deal I made in the final analysis with the boss was that if things were kind of messed up on the surface, then I wouldn't play with it because obviously we'd be accused of being too frivolous.

But if things had gone well, which they did, then the last thing I was going to do before climbing up the ladder to come home was to whack these golf balls, which I did, and I folded up the collapsible golf club and brought it back with me.

Shepard practiced with the special club—in his bulky spacesuit —while still on Earth, during the pre-flight isolation from earth germs.

"Given the stiffness of the suit, he could not hold the golf club with two hands," Pearlman said. "He couldn't bring his two hands close enough together to be able to hold the golf club and take the full swing. So he could only do it with one hand and he couldn't actually see the ball that he was swinging at because he couldn't look down far enough without falling over."

Nevertheless, Shepard was determined to try. The mission on the lunar surface was a success, and as Shepard and Mitchell loaded their equipment and prepared to climb into the lunar module for departure, Shepard paused, preparing to unveil his "experiment."

"Houston," Shepard alerted Mission Control, then "paused for effect," according to *Moon Shot*. When he continued, the Apollo 14 commander told all watchers they "might recognize what I have in my hand." He held the specially outfitted sample tool, which, he noted, "just so happens to have a genuine six iron on the bottom."

Back on Earth, NASA controllers gaped at what they were seeing. Shepard reached into a space suit pouch and held up a round object, noting, "In my left hand, I have a little white pellet that's familiar to millions of Americans."

"It's a golf ball!" a NASA controller yelled as grins flashed throughout Mission Control.

Shepard indicated he would attempt a one-handed swipe; there was "no way" to address the ball properly in his bulky suit.

"I'm trying a sand-trap shot," he told the world. He took an awkward swing. "I got more dirt than ball."

"Looked more like a slice to me," his fellow moonwalker Edgar Mitchell quipped.

Shepard then dropped a second ball, "determined to do better," he recalled.

And he did. The club made contact with the ball, which arched toward the horizon in the low gravity. "Beautiful," Shepard said. "There it goes! Miles and miles and miles."

Shepard and Mitchell boarded the lunar module and, after completing the preflight checklist, Shepard said, "Okay, Houston, the crew of Antares is leaving Fra Mauro Base." (The lunar module was named "Antares," after the bright star the astronauts used as a guide to their moon landing.)

And with that, they left the moon's surface. America's first man in space had now become the first man to golf on another world.

"The golf balls are still up there," Shepard said in his 1981 interview. "Perhaps the youngsters of today will go up and play golf with them sometime twenty-five or thirty years from now."

"WHY WOULD DADDY BE HITTING A GOLF BALL ON THE MOON?"

Except for the NASA officials who had to approve it, Shepard kept his moon golfing plans a secret. He didn't even tell his family. The secrecy was justified; if the mission on the moon hadn't gone well, he was not authorized to demonstrate golf.

"Daddy didn't tell us that he was going to be hitting a golf ball on the moon," Laura Shepard Churchley said. "I didn't know until I saw it on television. So, you know, Daddy was able to keep secrets from his own family.

"I was sitting up in bed, and I had dozed off," she said. "My grandfather, my father's father, granddaddy Shep, came over and nudged me and said, 'Laura, wake up. Your father's hitting a golf ball on the moon.' I thought something was wrong with Granddaddy Shep. Why would Daddy be hitting a golf ball on the moon? But sure enough, there it was. It was a surprise to me, you know? But Daddy was a golfer, it was kinda fun to watch."

Laura has a replica of the cleverly converted shaft and golf club head fashioned for her father's lunar golfing, and she demonstrated it during her interview, in Cape Canaveral, near the Kennedy Space Center. At the center, you can find wonderful Apollo 14 items on display, including Shepard's spacesuit, with lots of roomy pouches. (Designer Matt Radnofsky loved all kinds of pouches attached as pockets, a hallmark of his designs from sportswear to life rafts to space suits to the clever designs he sewed for his family. Matt made interesting jackets and gear, many with Velcro-attaching, removable pouches/pockets on large bags and clothes. The technique allowed for great flexibility for re-locating pockets, and easy access and stowage, depending on task and mission.)

Fellow astronaut Charlie Duke, who, as the lunar module pilot for Apollo 16, would become the youngest man to walk on the moon, also did not know of Shepard's "experiment."

"The golf part was a real surprise to me," said Duke, an avid golfer himself.

Golfing on the moon while millions of Earthlings watch on television is not the best way to avoid publicity. Shepard would later complain that he was better known for hitting a golf ball on the moon than he was known for being the first American in space.

"Alan Shepard came back as one of the most famous sports figures of all time," Pearlman said. "There was one list, I think Sports Illustrated, that rated him as one of the worst golfers in the history of golf, not taking into account where he was playing, but most people ranked him as one of the most famous examples of sports being played. With one simple sports act, he ranks in that list."

KICKING THE TIRES

Honored across the globe, Alan Shepard was to attain the U.S. Navy rank of rear admiral on December 1, 1971.

He knew his 1971 Apollo 14 moon exploration would likely be his last meaningful flight. Yet he stayed aboard at NASA and, with Slayton's help, got a job in the astronaut office.

The two friends worked together to select the astronauts for later lunar landings. And later, Shepard celebrated when Slayton, too, returned to flight status in time to train—and learn Russian—as part of the Apollo-Soyuz Test Project.

Each man "kicked the tires" for the other's journey, just as the old test pilots would have inspected their planes in their younger days. When Slayton watched his friend walking on the moon surface, he wrote he felt "one with Alan Shepard."

The devotion was mutual. Shepard considered retirement after his Apollo mission, but vowed to stay until Slayton could live his dream and finally get into space.

Watching Slayton take off to fulfill the joint mission with Russian cosmonauts, Shepard revealed he distanced himself from any technical aspects of the mission and instead pictured his friend Deke finally enjoying zero gravity; Shepard "knew it couldn't get any better for Deke Slayton," who had finally earned his astronaut's wings.

Neil Armstrong, the first man on the moon, wrote in the introduction to Moon Shot that Shepard and Slayton, used to working together since the early days of the Mercury program, expertly ran the NASA astronaut office, including the selection of flight crews. He lauded Alan Shepard in handling all day-to-day operations, making important personnel decisions, including design reviews and spacecraft testing.

A PILOT'S PILOT

What was the perspective and approach of Shepard toward carrying out a mission as a trained naval aviator and test pilot? We turned to another extraordinary pilot to see how expert pilots—particularly Navy test pilots—think. Shepard was a Navy pilot. It turns out this history has meaning in itself among the community of pilots.

One Monday at lunchtime at Carlos Beer Garden, we asked questions of a buddy of Carlos—former Air Force and NASA test pilot Terry Pappas, an accomplished writer and test pilot best known for his career piloting the SR-71 "Blackbird."

Pappas, a soft-spoken man with a modest, friendly demeanor, was dressed in golf clothes. Air Force Major General Pat Halloran, one of the few pilots to serve in both initial cadres of U-2 and SR-71 pilots, summarized the expertise of this quiet man at the beginning of Pappas's remarkable book:

> Terry has had some unique flying experiences in his many years in the Air Force flying trainers, bombers, and reconnaissance planes. In the Blackbird community, he is recognized as one of our very best pilots. He is a professional.

His performance on the many overseas operational missions he flew has been outstanding, despite challenges and emergencies he sometimes faced.

Following his SR-71 assignment, he was involved in diverse flying opportunities and spent many years as a research pilot with NASA, where he flew a variety of planes. These experiences have obviously added to his wealth of aviation knowledge.

Terry Pappas is a pilot's pilot. Astronaut Walt Cunningham, of Apollo 7, wrote about the SR-71—in Pappas's book *SR-71: The Blackbird*—with awe-filled reverence for the "wonderful challenge of controlling an airplane, flying at 80,000 feet and more than three times the speed of sound for hours at a time," to say nothing of the challenges of the frequent ballet of refueling in midair, requiring slowing to a near-stall, and falling out of the sky:

Higher, farther, faster—what every real aviator aspires to. The SR-71 was the epitome of this dream for three decades. The only way to beat the SR-71 was to rocket into space, and every astronaut in the office with me in the 1960s would have loved to fly the Blackbird. In many ways, it placed greater demand on piloting proficiency than any spacecraft.

We wanted to know: how would a skilled test pilot—Alan Shepard—approach solving problems and challenges, including the difficulties of demonstrating golf on the moon's surface—in conditions no human being had ever experienced?

The quiet man in golfing gear lent his calm dignity to the outside picnic tables at the Beer Garden as he explained what would distinguish Navy pilots, including Alan Shepard, from other pilots. He began with an emphasis on Shepard's training and background: focus on the mission.

I find that Navy pilots tend to have a broader perspective than typical Air Force pilots. The ship comes first to the Navy and airplanes come second. The Navy will push an airplane over the side of the ship if they think it's necessary. Air Force pilots are trained to dot every i and cross every t. And they will be punished severely for mistakes which cause the loss of an aircraft.

The Navy trains pilots a different way. Do what you have to in order to get the job done. It's somewhat broader. It's completely different. A completely different approach.

A remarkable lady from South Africa having lunch across the Beer Garden picnic table jumped in with a cogent, key question. Her name: Sally Antrobus. (We learned only later Sally was a highly respected book editor and an integral part of the NASA community. She later served brilliantly in key roles editing for this book.) At the lunch table, she was sandwiched between a physician, a Rotary president—now a Rotary Governor—and a current local city councilman. Sally asked what accounted for the different views between the Air Force and the Navy test pilots.

"The Air Corps—now called the Air Force—doesn't have the history," Pappas said. "The Navy, however, is as old as the Army. The Air Force is still new and upcoming. They still have to dot the i's and cross the t's. In the Navy, the ship is the mission, and you look to the best interest of the ship; it's your city on the water. You do what you need to do to carry out, protect, your mission. "

THE YOUNGEST MAN ON THE MOON

Apollo 16 astronaut Charlie Duke, the famous, trusted, calm voice preferred as "CapCom" (Capsule Communicator) by multiple Apollo astronauts, vividly recalls Alan Shepard half a century later.

The two men were also friends who enjoyed playing golf and, particularly, playing golf with legends of golf, childhood heroes.

When Barbara asked Duke to introduce himself, he first spoke movingly of Matt Radnofsky as a great friend to the astronaut corps. He remembered the people of Crew Systems behind the scenes and reminisced. The documentarian at the first camera had to squeeze hard on Barbara's shoulder to remind her to stop crying happy tears and get on with the interview as Ed kept calm, and carried on, filming on a second camera.

> I was an astronaut from '66 to '76, participated in five of the nine missions that went to the moon. I was CapCom on Apollo 10, CapCom on Apollo 11, back up on Apollo 13, flew on Apollo 16, and then back up on 17, was the 10th man to walk on the moon with John Young when our group, a group of 19, arrived in April of 1966.

He sat back, looking handsome and youthful, fifty years after his landing on the moon.

"And you were the youngest man to do so," Barbara added.

He paused and smiled broadly, with a twinkle in his eye. Then he said:

"And I still am."

He immediately and enthusiastically shifted to talk about the Artemis program, NASA's next step in human space exploration, to establish a sustainable presence on the moon to prepare for travel to Mars.

He hopes and expects, he said, a woman will soon replace him as the youngest person to land and explore the moon.

Duke graduated from the U.S. Naval Academy at Annapolis and chose the Air Force for greater opportunities for his flying career. He was the calm Texas voice of Mission Control the astronauts wanted for their communications. He and John Young are renowned for their historic geological discoveries in their lengthy Apollo 16 moon explorations.

The astronaut spent much time with us conveying his great admiration and respect for Alan Shepard, getting to know and work with him when Shepard headed the astronaut office:

> He was under a lot of stress at this point. I think before we got there, there had been three astronauts killed in airplane accidents. Ted Freeman was killed, a T-38 crash at Ellington. Two guys were killed up at McDonnell, Elliott See and [Charles Bassett, prime crew, Gemini IX]. And so it was a tough time.
>
> And right after we got there, Ed Givens [Naval Academy graduate, USAF officer, test and fighter pilot chosen for the Fifth Astronaut Group] was killed in an automobile accident. And in the summer of '66, Alan was grounded. And it was probably a frustrating time for him because of his medical condition.
>
> And so as I got to know him, I really respected him. He was very knowledgeable and sharp, but tough.
>
> He wasn't going to take any grief from anybody, but I like that; you knew where you stood. His secretary—I believe her name was Gale—if he was in a good mood, she had a smiley face on his door. If he was in a bad mood, it was a frown. And so most people knew not to knock on the door and go in his office at that point.
>
> But he ran a good office. He was chief of the astronaut office, and I really respected him after we'd been there about nine months, and we were beginning—not to get flight assignments—but we were beginning to train in geology, and they kept us busy.
>
> We all had extra activities to do; my job was to follow the development of the lunar module engines and the service module engine and also the Saturn.

Duke explained just how important Alan Shepard was as a great mentor and colleague. He became quiet and serious as he described Shepard working alongside the then grounded-for-medical-reasons Deke Slayton.

"WE'RE NOT GOING TO GIVE UP ON THIS"

Duke spoke of all the Apollo astronauts' serious responsibilities on the ground to prepare for the manned moon mission, particularly after the tragedy the country experienced with the horrors of a fire in the capsule cabin during testing on February 21, 1967.

The fire killed all three crew members: Gus Grissom, Ed White, and Roger Chaffee. NASA designated the Mission as "Apollo 1." Duke calmly reflected on the important leadership Alan Shepard provided during this difficult key time, referencing the astronauts' responsibilities:

Stu [Roosa, Command Module Pilot of Apollo 14], Bruce [McCandless, support crew of Apollo 14 and the first astronaut to fly untethered from his spacecraft], and I were assigned to monitor the development of the Saturn V and so we'd report back once a month. We'd go up to, for instance, we'd go up to Huntsville or to Marshall, where for Von Braun's meetings we'd sit in and just listen to the development. Then we'd go back and report to the astronaut office what we thought the schedule was and how it was developing then.

So Alan was monitoring. He and Deke Slayton ran the astronaut Monday morning meetings, if you will. And so then … in January '67, we hadn't been there quite a year when the fire occurred. And that was a big shock, of course, to the system and to us personally. Yeah, I couldn't believe it.

It added a lot more stress to the two guys like Slayton

and Shepard, I thought, and to everybody, you know, we're dead in the water. How are we going to make it to land on the moon by the end of 1969?

Duke told us that "Alan was a good motivator" to keep them pressing on. The youngest person to date to walk the moon quoted Shepard's reassurances to his team: "We're not going to give up on this, guys."

Duke provided fascinating detail on this historic time, as the astronauts were deeply engaged:

> And so we all had various activities as a result of the [deadly Apollo 1] fire. I remember Roosa and I were assigned to the Emergency Egress Working Group. Once we get this spacecraft fixed, how are we going to get out of it? And what do we do if and when we get out—do we go down a slide well or do we go in the elevator down? So we work with this team and then we report back on Monday morning.

Duke used the same word as Shepard's daughter had used to describe her "stern," yet loving and sentimental dad: "Al, I never I never saw him really angry, but he was stern. He didn't take any lip." The astronaut sitting across from Barbara smiled thoughtfully —more than fifty years later—as if reminiscing about a brother who has passed:

> I never had the courage to argue with him.
> I had nothing to argue with him about, actually. I felt he ran a great office and—I don't know when he actually got back on flight status—but then he got assigned to Apollo 14, really Apollo 13. But they swapped the crews, and he became Apollo 14, with Stu Roosa and Ed Mitchell, and he changed.

Then he dropped out of the astronaut office. The head of the astronaut office, if I recall, and began training and was back on flying status. He was a real pleasant personality and so I enjoyed him.

Duke spoke about Shepard's trustworthiness and calm, stern approach toward serious astronaut responsibilities, trusting his colleagues. The Naval Academy graduate and Air Force pilot explained how he felt as an astronaut under Shepard's strong leadership as a boss who had his back when Duke took strong, principled positions for science and safety:

You're grown, man. You got selected as an astronaut. And you ought to be able to work on your own. With the additional duties I had, I would report to him. I worked on monitoring engine development, but we had a problem with the ascent engine and the lunar module, and it was uncertain—not unstable, but marginal .... So Alan, Deke, and all of us got concerned about it.

George Lowe, who was program manager at the time, formed this committee to go decide whether we're going to keep this engine or we're going to get another one. I think he gave us four or five months. I was on that. I got put on that committee—only astronaut on that committee.

We went to who was making the engine. You know, not everybody was making rocket engines back in those days .... And so we went to them and listened to their proposals and what they could do. And it took us about six months, but we went back and, of course, Alan and others were very interested. And then I said, in a pilots' meeting, I said, you know, I think we need to change contractors and so when we got to brief Lowe, he went around the room and personally asked each of us, what did you think?

I think I was the last guy. And I said, I think we need to

change, we change contractors. Well, the basic engine stayed the same. It was just the injector plate. It was causing the problem. And [we needed to] break it down, solve that problem.

So — we had a very stable engine for the lunar module.

AN ATMOSPHERE OF RESPECT

Duke noted Shepard's fairness and trustworthiness and his calm: "Alan—in all of our pilots' meetings—I never heard him jump up and down and stomp and scream at anybody or anything. He was very cool. Very calm, stern, as I said earlier. But you knew where you stood with him. I mean, if he didn't like the way things were developing, you'd have a discussion. I thought he was very fair—and you know: everybody is on a first name basis."

Duke emphasized that the relationship was much more than boss-employee; the astronauts could confide in—and trust—Shepard: "He's your boss. But hey, 'Al, I got this problem,' you know, and he monitored our military records also … you were actually under Alan Shepard as your boss because he's the senior guy in the astronaut office."

Duke also spoke movingly of the importance of the entire team of workers; NASA's astronauts respected all levels of workers. He told us Shepard treated with great respect the astronauts with whom he worked. And when the iconic, sidelined "first American in space" regained his medical qualification to finish his astronaut service as commander of Apollo 14's mission to the moon, Duke told us, Shepard was a first class colleague.

Duke emphasized that Shepard created an atmosphere of respect that extended to all workers.

And then our two suit techs were real close. We were real good friends. And so you just sort of relaxed around them. You know, they had your best interests at heart. And so you got to be more than just [helpers]; hey, he's not just my suit. He's my buddy, you know?

And so you trusted him that this suit was going to be ready to go when you needed it and in training and whatever we had. I was support crew for Apollo 10 and Cooper was the backup commander on that flight. So I got to know him working with him in procedures and all.

You could trust them, and they gave you a job .... And once you got a crew, then the commander became more important than Al. But I found working in the astronaut office was a real delight. I really enjoyed getting to be buddies and comrades with these guys. And they were all hard workers, talented; you had confidence in everybody [to] make the right decision.

Duke gave insight into the complex, life and death world of the astronauts, and the role of Shepard's leadership skills in ensuring a sense of equality.

When we had problems like airplane accidents or when C. C. Williams was killed later on a T-38—and Pete Conrad ejected from a T-38—that got everybody's attention, you know, and people wanted to find out why. And so we all worked together on those kind of things and everybody had real concern—it was more of a fraternity.

And that was hierarchy level stuff. You know, everybody—even though Alan was a captain in the Navy, I was a captain in the Air Force—like three ranks below nobody. You knew he was the boss, but you didn't go in and say, you know, Captain Duke reporting, sir, you just walked in and said, "Hey, Al, I got this problem." And so it was that kind of attitude. Yeah. You were sort of equal.

A COMMITMENT TO SCIENCE AND EDUCATION

Duke reflected on Shepard's charitable works, his devotion to education, and his role in forming the Astronaut Scholarship Fund at a local bank. The fund brings together the resources of the original Seven Mercury astronauts after Gus Grissom's death in the Apollo 1 fire. It was his commitment to education that led Shepard to chose an experiment that would appeal to nearly everyone: using golf to demonstrate gravity and physics.

Duke explained that although Shepard preferred privacy, he used his fame for nonprofit scholarships and to help people and causes he championed. The leadership of the internationally respected Astronaut Scholarship Foundation* includes both Charlie Duke and Laura Shepard Churchley.

They carry on Shepard's educational legacy, ensuring the United States will "maintain its leadership in science and technology by supporting some of the very best science, technology, engineering, and math college students."

After hearing Duke's explanation, we now support this organization, and Ed is a volunteer mentor.

"AL RADIATED CONFIDENCE"

As a young lieutenant, Duke recalls admiring Shepard as the "first guy in space," whom he recognized as a talented astronaut, then as a level-headed leader motivating the astronaut corps. Then, once Shepard was back on flight status, "we became just a crew together." And Duke loved playing golf with his childhood heroes and Alan Shepard.

Duke reminisced about being named an honorary member at Champions Golf Club, on Houston's north side, with Shepard and fellow Apollo astronaut Gene Cernan. Excited as a schoolboy,

---

*   https://astronautscholarship.org/scholarshipprogram.html

Duke told us: "And so we got to meet Jimmy Merritt, who was one of my heroes as a boy. He was a great golfer during my boyhood and so was Jackie Burke. And so I got to play with them! I got to know them."

Duke recalled: "And—we were in the race again, if you will. The Russians were still ahead of us at that time with Yuri Gagarin." Of Shepard, Duke said: "Al radiated confidence. And authority; you knew who was the boss in the astronaut office, and so you—I think [everyone, and] I personally — would really be motivated by him. I saw him as the leader. I wanted to do a good job to support him and to support the astronaut office ... and my role; so he was a good motivator for me. And once he got back, as I said, once he got back on flight status, we became just a crew together."

Duke's reflections were consistent with the explanations we received from Laura Shepard Churchley. After what was a difficult time for him personally, as his daughter explained, Alan Shepard underwent a new type of surgery that put him back into the lineup. With intensive hard work, he went on to command Apollo 14 in 1971 and became the fifth person in world history to walk on the moon.

A SECRET 'TIL THE END

The golf-playing Duke—who shared Shepard's love of the game and of golfing with their sports heroes—noted of the Apollo 14 mission that Shepard kept his lunar golf plans quiet, even among his fellow astronauts:

> We got a big setback from 13 when the oxygen tank blew up, so we were all involved in that and integrating the knowledge from each flight into your procedures and into your training, and I don't ever remember him mentioning in one of our weekly meetings that he was going to hit a golf ball on the moon.

And if I remember, [the golf club head] fit onto this handle of this shovel; we had a handle that you could take the shovel off, put this little club on, and then swing it. It was a big shock, but I thought [it] was a great idea—at the end of his time on the moon—to celebrate with a game that he really loved.

A HEADLINE-MAKING HAIRSTYLE

As a public figure, Shepard cared about his appearance. In his 1981 interview, he explained that his admiration of the many people who enabled the success of the space program influenced his own sense of obligation to maintain a positive public image.

While Shepard ensured his golfing plan was never leaked or publicized until it happened, the press was continuously hungry for interesting stories during the flight and as the world waited for the astronauts to be released from post-flight quarantine.

Carlos was included in some of the astronaut "human interest" stories, which focused on Shepard's care about his hairstyling and appearance.

During Apollo 14, the press and their readership followed every aspect of Shepard they could find, including his appearance, his hairstyles, and his barber. National press interviewed Carlos in his barbershop, discussing his haircuts for the handsome astronaut. The thirsty news media focused on Shepard's prior hairstyle changes, as well as his frequent trips from the Cape to Carlos's barbershop for maintenance of Shepard's change from a crew cut to a modern hairstyle, as he switched from a military crew cut to a 1970s "Shepard Shag."

The press reported on Shepard's style changes and his haircuts, including a scheduled haircut after the moon return, per press reports of interviews with Carlos. The Shepard haircut the night before isolation was well-reported in U.S. and foreign papers, as part of a flurry of human-interest articles surrounding the mission, its successes, and the returning astronauts soon to be on display.

Lee Holley, popular local newswoman for the NASA-area News Citizen, reported details in a piece titled: "Shepard's Hairdresser Knows His Astro-follicles," in the physical copy of the newspaper that Carlos and his family had "decoupaged" in the sixties style, mounted with glue on a piece of wood. You could still find it on the wall of the barbershop fifty years later.

Holley used two photos to illustrate her "Astro-follicles" article, comparing Shepard in his old crew cut with a second portrait of Shepard with his recent "Shepard Shag" or "businessman's haircut." She wrote:

> Carlos went to Shepard's plush home ... the Sunday before he began the pre-mission twenty-one-day quarantine, to give him the last haircut until he comes back from the moon. He said he cut his hair in the living room right on the carpet, as his wife Louise watched, fascinated. She said she had never seen her husband get his hair cut.

The longtime local newswoman understood the delayed haircut because of two quarantines and the flight in between:

> Shepard usually gets his haircut every three weeks, when he's in town, but the twenty-one days pre-mission quarantine added to the nine-day flight, and the twenty-one days quarantine after the mission means it will be about fifty-one days. That is, unless they decide to let Carlos go into quarantine quarters at the Lunar Receiving Lab, which is probably unlikely.

The local reporter explained the virtues of Alan Shepard's hair —and the stylist:

Carlos can vouch that even though he's forty-seven, Shepard has no gray hair (and doesn't dye it either) but dark brown hair, that tends to be as stubborn (as he is). That is why it is combed forward, because it's easier to manage. Carlos said he told Shepard not to forget his brush on his trip to the moon and is always bugging him to brush his hair more. He said Shepard quipped, 'Hope I don't forget my brush and spray,' but Carlos says he doesn't use spray. Most of his other customers do, but he prefers to call it hair trainer.

Carlos is a popular hair stylist in the area, and in addition to astronauts, cuts hair for many of the Space Center Rotarians, and NASA and aerospace engineers.

Holley provided more style pronouncements, noting, "The flat top is out," while the new trend in the 1970s will be "new shag, plus growing moustaches and beards."

In the interview, Carlos reveals the trend of long hair for boys in school, noting he had been polling his customers on the issue. Most of the customers—and some mothers of schoolboys—favored letting students make their own choices.

The Dallas Morning News reported—in a clipping Carlos's loved ones mounted onto a wooden board in the same style as the "Astro-follicles" piece—on the charm behind the astronaut during the home visit for a last trim: "Despite training pressures, Shepard took Villagomez on a tour of his house, since the last trim took place in the astronaut's living room."

Carlos described his visit to Shepard's house to give him a haircut before the the mission: "Visiting Shepard's house last month before the Apollo crew went into preflight quarantine, Villagomez cut the astronaut's dark brown hair short enough to last forty-five days."*

---

* *The Dallas Morning News,* February 10, 1971, 16A

Carlos told us:

When I went over to his house, he says, "Carlos, cut it off."
Now, somebody else might have just taken a clip. But I
styled it a little bit. It was only that long. Then shortened
the side, and I think that's what he liked.

What I was going to do to him, he was not going to like,
but he said short. So I weighed it and I cut it a little less
than the thickness of my fingers. I just picked it up and
then I got my clippers and I cut it. It was almost a burr;
almost. But it was a—it was a style burr. You know what I
mean? Had a little something. A lot of people [would] just
cut it. Don't care, you know?[*]

But Carlos said he made "sure it had a line here and a line
around the hair and [looked] a lot better. And he liked it."

Carlos cared—even more than his friend and client—that Shepard's shorter cut for space was as good as possible. It made no difference to Carlos that no one would see Alan Shepard's hair; in
talking about barbering, Carlos exhibits the manner of a consummate professional, a perfectionist in haircutting. Carlos cared
deeply that this haircut was done properly.

The Dallas Morning News then published a "Staff Special"
piece on "Shepard's Cut" and a "Shag Look" after splashdown, on
the heels of a United Press story picked up nationally and internationally. The Dallas paper added news from their in-person interview with Carlos, informing their readers that Carlos had watched
the splashdown the day before (February 9) and that Alan Shepard
"will be sitting in Villagomez' barber chair less than three weeks
from now when lunar quarantine was lifted"[†]

Shepard's pre-mission haircut by Carlos built on an intense

---

[*] Villagomez interview, April 15, 2022.
[†] *The Dallas Morning News*, February 10, 1971, 16A

year of experience in developing a new hairstyle for Shepard. The two men had worked together seamlessly to change his "flat top" military crew cut to the "banker's haircut." A key element was maintaining privacy and dignity, which included no photography of the styling sessions.

After experimentation with different styles, Carlos and Shepard achieved a new look. Carlos kept his promise to his famous friend that there would be no need to go through the "ugly stage" other military men faced when attempting to grow out their crew cuts.

When NASA press man Gene Horton recruited Carlos, he wanted a barber who would be skilled in helping military men move on from their "flat tops."

To his credit, Shepard maintained the discipline needed to follow Carlos' directions during the difficult transition period, tolerating the use of a goopy product under a hairnet, baking under a women's hair dryer, and the use of hair spray.

"And don't call it hair spray," Shepard cautioned Carlos.

Carlos described the arduous trial-and-error process resulting in the haircut then famously known as the Shepard Shag. It was all built on friendship, loyalty, trust, and confidence—and the fact that only Carlos could do the job properly and achieve the handsome haircut and styling.

"Shep was having a rough battle," Carlos said. "I said, 'okay, Shepard, you don't want to go through the ugly stages,'" Carlos said.

As Shepard prepared for the drying under-the-ladies-bubble-dryer part of the styling processes. he told Carlos: "You know, I don't want anybody coming in here."

Once Carlos and Shepard had experimented, tested their procedures, and achieved the desired result, Carlos promised an efficient process. He knew Shepard did not want to be photographed under that dryer. He promised Shepard he could complete the process in five days, and then he told him:

You have to be here every morning about eight. We open at nine. It only takes fifteen minutes of your time." Shepard agreed .... He came over and I shampooed his hair and I put a lot of gel and then I got a hairnet and I put it around him and tied it up.

Shepard's big concern remained avoiding photographs.

So here we are, flat, you know. And I put him under a hairdryer. Boy, he didn't like that. Well, it was all "make sure they keep the doors locked. Otherwise I'll always be under here and [they'll] take a picture of me." But it only took like five minutes to dry his hair.

The process continued with a stiff brush, rather than a comb. And hairspray.

Then, when I took the net off, the hair is down. So you take a brush, then you break it up and it stays down. It comes up a little bit. Then I put a little bit of spray and he said, "Don't use the word spray."

He was pretty, pretty particular. So he was impressed that the hair was not sticking straight up.

And when trouble did come, Carlos was right there to solve the problem privately, as he told us in the interview:

But he called me the next morning. He said, "When I came out of the shower, my hair's sticking up all over the place." Well, I say, okay. Come back tomorrow morning. The next day, come back. We did that five days in a row and by the fifth day, you stay down. Okay, at that time, he had a nice haircut.*

Carlos told *The Dallas Morning News* that Shepard relied on him "because he wanted his flat top changed to long hair"[†] And Carlos explained their scientific method: "We tried three or four different styles. With his hair parted, he looked too conservative. He's not a conservative guy at all. Finally, we let it all grow an inch and a half long, cut and dried it, and just let it fly."

Carlos also told *The Dallas Morning News* that Shepard "has no gray hair." The newspaper further reported: "The style has become popular here and Villagomez hopes to see it tagged 'the Alan Shepard cut.'" To this day, Carlos tries to tamp down the use of the word "shag," adamant that the word does not adequately describe the more conservative style Shepard chose.

"It was not shaggy," Carlos insists whenever someone asks him about the Shepard Shag. "It's a conservative banker's haircut or businessman's haircut," for that time. But the press phrase had stuck.

The *San Francisco Chronicle* reported on Shepard's practices for getting haircuts after the change to the longer style; he would fly back "from Cape Kennedy to Carlos for a six-dollar haircut ever since he decided on the new, over-the-forehead style. The new style is razor cut, 1½ inches long all over his head. It doesn't need combing"[‡] The California newspaper also repeated Carlos's praise for his client's original hair and color: "He has a very full head of hair for a 47-year-old man. A lot of people think he dyed it, but he didn't."

The date of the *San Francisco Chronicle* coverage, Friday February 5, 1971, was the date of the Apollo 14 lunar landing, and the piece explained that "Shepard paid Carlos fifty dollars a year

---

* Villagomez, interview, April 14, 2022
† *The Dallas Morning News,* February 10, 1971, p. 16
‡ "Shepard's Hunt for a Hairdo," *San Francisco Chronicle,* February 5, 1971, p. 9

ago to find him a new style to replace the crew cut he had worn since before his first space flight ten years before." They quoted Carlos:

> He'd been wanting to do it for a long time but didn't want to go through that bad state. We tried three–four styles. We left it flat on top, but it stuck straight out. It's stubborn hair. We let it grow a lot longer, but that was no good because he couldn't control it. He's always wearing helmets. We put a part in it, but that made him look sort of strange .... It made him look more youthful. The man, close up, doesn't look thirty.

The Los Angeles Times ran the same story, also on February 5, repeating some of the same United Press information, as did the *Washington Post* on February 6, quoting Carlos complimenting Shepard for being conscious of his hair. The *Times* and *Post* ran variations of similar details, as did other papers, quoting the younger Carlos on Shepard: "He's a cat, man. He drives real fast cars. He dresses sharp. He's an up-to-date fellow. He's got a great sense of humor. He's always got a nice neat joke, up to date."

A LOVE OF FAST CARS

Carlos told a story in our interview about Shepard, who did indeed enjoy fast cars, as the newspapers reported:

> One day, by the way, he drove up here, and he called me, says, "Can you get away from the shop for about an hour?" I said, "Sure, I think I can, will be ready." And he pulls up here in an English-looking car with three seats in front, nothing in the back.
>   I mean, it was a sporty-looking thing. So we go down highways. He had to go to Texas City or something and

pass Dickinson. There's a wide open space on Highway 3, and he went so fast, he was doing like 125 miles an hour.

"And, man, I don't like speed. That's Shepard. Slow down, man. We're going to roll. You know, he started telling me about aerodynamics. He says, 'These cars are built to go fast.'"

Carlos worried the car would fly off the road, but Shepard calmly assured him that the car would be fine.

"The wind's pushing it down," he said. "Well, what if a cow is in the middle of the road?"I said. He says, "Carlos, that's not supposed to be there." That was his opinion. I was never so happy as when that thing slowed down.

THE "SMARTEST BUSINESSMAN" OF THE MERCURY 7

NASA Public Affairs Officer Gene Horton's memoir notes Shepard had a reputation for well chosen, considered investments, including in stocks and banking. Horton wrote that the astronaut profited with smart stock market trades "and developed a reputation as the smartest businessmen of the Original Seven."

Carlos told us that Shepard was constantly curious, asking important questions in the NASA area about land, business, and finance, asking Carlos to be his eyes and ears in the community. Carlos was a natural leader with a respected reputation as a local businessman with entrepreneurial spirit—as a lifelong Rotarian, former city councilman, and mayor pro tem, and the owner of a still-thriving Beer Garden next to his barbershop. He remains a pillar of the Webster community in religious, charitable, business, and governance matters.

He told us he was well aware of Shepard's quiet successes in banking, real estate, and related developments; Shepard always had a keen interest in understanding the NASA area and its business potential. The astronaut knew, too, how to take advice and exit gracefully. He enjoyed success as a beer distributor after his NASA career, relying on Carlos for insights into the beer business—especially as the astronaut learned to sell Colorado beer in the Budweiser country of rural Texas.

Shepard's daughter, Laura, said that after her father left NASA, he owned a Coors distributorship with one of his Navy buddies, Duke Windsor, a good friend from their test pilot days who was married to a Coors family heiress. The old Navy buddies had both achieved the rank of admiral.

Although she had never met Carlos, Laura Shepard Churchley knew of her dad's friendship with him: "Daddy was a good judge of character in people and he probably saw a trusting person in Carlos. I mean, how often does a man get a haircut? Maybe every two to three weeks. So they got to know each other very well, I'm sure."

The astronaut with the cool demeanor could indeed make and keep warm friendships. As a girl, Laura was aware of her dad's life at NASA and in the Webster community surrounding the Space Center, and of his comfort with Carlos, as her dad's haircuts merited dinnertime conversation. She told us:

> I know they called him—some astronaut referred to him as the "ice" commander. And so there were times, I'm sure, that he could have been a little more understanding. Oh, but he learned to deal with it.
>
> And as we all do, we all knew that Daddy had a barber named Carlos. Yes, we know that Daddy had a barber in Webster, Texas. I'm sure he was comfortable with [Carlos] because he stayed with him for so many years.
>
> And because we would notice at the dining room table that he had a haircut, we would make a comment: You got your hair cut. Did Carlos cut your hair? And yes, we all knew about Carlos, but we never met him.

BEYOND HAIRSTYLES

From the time we were children, we knew from early visits to Carlos of his relationship with Alan Shepard. Carlos still proudly displays the aging, extravagantly signed posters, and photos from the exciting early days, including one big poster with double inscriptions from Shepard. His friend added language as time passed, with the astronaut signing proudly with reference to a beloved car and a license plate.

A handsome Villagomez family snapshot was taken with Shepard, both men, years younger, posing with Carlos's family, dressed in early 1970s attire—with an equally nattily dressed Shepard.

Carlos's friendship with Shepard went well beyond fashioning a hairstyle to work for his friend's difficult hair, of course. Shepard shared his barber's entrepreneurial spirit, and he frequently asked Carlos questions about the community. It was a community that Carlos helped build through many acts of charity—donations of food, medical care, transportation, and dozens of hand-crafted benches auctioned off and still on display throughout Webster. One former Rotary president said that Carlos was "as important a figure as could be found" in the long history of the local Rotary.

THE ENTREPRENEURIAL ASTRONAUT

Carlos taught Shepard more than how to handle himself with bar workers and ranchers and day hands. He also helped the astronaut understand the culture as they traveled into bars around Texas after Shepard acquired his Coors distributorship. The once-reticent airman learned from Carlos the art of talking with rural bar owners and customers.

"I already had the Beer Garden," Carlos said. "So as soon as they got their franchise, Shepherd came straight to me because he had to learn how to sell beer. And he didn't know nothing about beer."

Shepard and Windsor also tried, in their own ways, to help Carlos. He shakes his head as he tells the story of Shepard and Windsor paying a visit to his home one afternoon. The friends trooped upstairs to where Carlos was recovering in bed from a rather delicate male surgery. The two retired admirals watched children's cartoons on TV, joking, laughing, and trying to keep up Carlos's spirits.

Carlos was hurting. He wasn't feeling much like laughing. He remembers the men trying to distract him—and his lack of cooperation with the distraction as the men roared at the cartoons. Carlos still views with wonder the memory of the two admirals cutting up as the bosun's mate, third class, looked on dourly in his discomfort.

Shepard consistently asked Carlos to go with him on gliding trips out west. As a Navy man, Carlos was experienced with boats and ships; he rejected, however, all invitations for air travel with Shepard as the dazzling pilot.

Prior to Shepard's retirement, Carlos says his client was heavily focused on his NASA work, competitive with his fellow astronauts, and genuinely interested in the greater NASA community. Shepard particularly sought information about Carlos's Mexican heritage, culture, and language. He wanted to learn Spanish, so Carlos taught him.

Shepard always sought to learn, and Carlos was an excellent teacher. Carlos's life story is a story of great spirit, toughness, kindness, risk-taking, intelligence, and service, including his honorable and enduring devotion to his family, community, and country.

No one, not even the first American in space, could have a better friend than Carlos Villagomez.

CHAPTER 2
THE BARBER

ON THE DAY NASA ended the Apollo 14 mission quarantine, Carlos received an expected and welcome phone call from "across the street"—the NASA Manned Spacecraft Center: "Mr. Shepard's coming down for his appointment."

The barber stepped outside to the parking lot and quietly stood on the pavement. His barbershop remained quietly housed in a Holiday Inn hotel room, which Carlos had rented at the urging of his NASA press friend, Gene Horton.

Carlos awaited his most famous client, just as he'd promised seven weeks earlier. A car pulled into the parking lot and a joyous-looking Alan and Louise Shepard emerged, keeping Alan's scheduled hair appointment in Room 105, which had direct access from the lot.

Carlos vividly recalls the smiling couple holding hands and extending warm, elated greetings. Shepard was dressed in a flight suit, the messier, drab jumpsuit coveralls based on those used by hard-working military crews in World War II—not the more tailored, handsome, blue suits dressed with patches and designed for photo opportunities.

Carlos recalled Shepard looked a happy mess, including

jumbled and unkempt hair after more than seven weeks of growth during pre- and post-flight quarantines and space travel. Carlos said it looked "like a firecracker had gone off!"

The Shepards' appearance called to Carlos's mind the extraordinary evening with them weeks earlier, when he had cut Shepard's hair the night before isolation as Louise watched.

Carlos had "cut it short," recalling Shepard's reminder that he'd be gone a long time and would need a haircut when he got out of post-mission isolation: "But I want you to be available .... I'll come straight to your shop."

AN EXTRAORDINARY FAMILY

What brought a man from the segregated wards of the Houston Ship Channel—from a family of fifteen children—to sharing drinks and musing under the moon and stars with the first American in space, on the eve of the astronaut's next U.S. mission, to pilot a lunar landing and to walk on the surface of the moon?

Carlos was part of an unusual, disciplined family born to hard-working parents who also raised others' children in their home. They believed that "everyone needs a trade." His parents suggested two possible occupations for Carlos: barber or cobbler. While Carlos is also a skilled shoemaker, everyone, including Carlos himself, favored barbering. He made the big choice in consultation with his dad when he came home from Navy service.

The Villagomez family made their home during the early part of the twentieth century in Magnolia Park, a neighborhood where World War I émigrés from Mexico and their children formed a tight-knit community centered on family and heritage. Papá Villagomez worked hard at a large cement company for forty-four years. Carlos was born between the world wars, on May 20, 1936. He recalls that his mother "spent her life in the kitchen and was always feeding a baby."

Carlos's family lost one child, Margarita, to tuberculosis. They raised more than their own thirteen surviving children. In hard times, especially in the time of rampant tuberculosis, the family took in two more children permanently. These youngsters, too, took the Villagomez name. Carlos stays in touch with them and their families. After all, he is their big brother.

The Villagomez home, with one bathroom, housed an entire neighborhood of family. The boys used an outdoor shower and privy, which had hot and cold water after Papá got large water heaters from the old Ellington Air Force Base. When the children learned English, "Papá" became "Dad." Their mother never learned English, so she remained "Mamá." Carlos remembers Mamá telling him, "Your Papá is working himself to death so we can be fed."

Papá built an apartment over the garage. The small, safe, private area housed folks who had nowhere else to go. Over the years, the Villagomez home and apartment served as a refuge for many people in conflict, several of Carlos's grown sisters with their babies, and the children of others in need. Carlos lived there himself for a bit as he decided on a profession after he returned from distinguished Navy service on the other side of the world.

"I LOVE THE NAVY"

All the Villagomez boys enlisted in the services as soon as they got out of school. Carlos chose the Navy. Proud of his service, he says he enjoyed working with ships and handling boats, including his captain's gig, a sign of the captain's trust in him.

Young Carlos called home to tell Mamá she should have all the boys choose the Navy for enlistment. "Tell them," he insisted, "tell them all. I love the Navy. I love the food, there's always enough to eat, tell the boys to choose Navy. The food is great!"

From 1953 to 1957, Carlos served on a dock landing warship, the USS Comstock LSD-19, in Indochina during the Korean conflict. (A dock landing warship is a Navy vessel with the ability to transport and launch other vessels, such as landing craft and amphibious vehicles. Carlos describes the noise of bullets hitting their warship and its gate, which would close as refugees rushed aboard under fire and then finally protected from the hail of bullets after the gate closed.)

In Korea, the ship's missions included evacuation of U.S. soldiers to the Japanese coast. Carlos first developed painful ringing in the ears from the noisy enemy fire that rained down on them as the soldiers ran aboard. He learned to live with it, and still does.

The warship carried equipment, Marines, and Navy frogmen, approaching land in routinely rough seas to launch them on their missions. Carlos marveled at how the frogmen jumped from the ship as it traveled at speeds of fifteen to twenty knots.

These predecessors to the modern SEALS encountered twenty-foot waves, made the arduous swim to shore, conducted their missions, and calmly returned at the appointed hour. Those were the men Carlos admired, although he never knew the details of the frogmen's missions. These men told Carlos that it was all in a day's work and something "you just get used to."

In what is now Vietnam, the ship's missions included moving the fleeing Catholic population from the north to the south. The ship and its crew saved thousands of people, including a physicist who walked into the barbershop years later. The scientist realized Carlos had been a sailor on the ship that had saved what remained of his family. The man asked Carlos to stay late, please, at the shop. To Carlos's surprise, the customer returned later that evening with his sister and aged, blind mother, who brought a most treasured photo with them. It was a picture of the ship that had saved their lives.

The father of the family had disappeared and never made it aboard the warship; this now-grown man and his family wanted Carlos to know the difference the U.S. Navy and his service had made in their lives. The little boy had become a scientist, his older brother a respected veterinarian, and his big sister a fine teacher.

Carlos also volunteered as a test subject to determine the effects of dropping an atomic bomb during tests at the Enewetak Atoll in the Marshall Islands. He recalled being told that the Navy wanted to test what would happen to a ship and its occupants within two to three miles of an atomic explosion.

He says the effects were "spectacular." That old test ship shook, pipes broke, the blast waves and wake were huge, and the scope and effect of the waterspout and stem of the mushroom cloud were even more extraordinary. Eventually, Carlos's test ship sank. He received a certificate for his service as a volunteer subject. Years later, the Navy followed up with him, asked about his experiences during this period in Indochina, and offered him partial disability, they said, for his lifelong ear problems.

Carlos finished his service and found his way home to Houston, to the wife he'd married the year prior on leave: "I had got married before we took that last cruise. And when I came back, my wife had a baby. So I had not seen my wife in about eleven months. And then when I saw her at the airport, she looked different because she had had a baby and it was kind of exciting."

BECOMING A BARBER

With input from his parents, he decided on a profession. As noted, he was a talented cobbler, but he "took about five or ten days to just hang out, see what I was going to do. And that's when I decided between cobbler and, yeah, the barber."

The Navy paid for his training under the G.I. Bill. "They give you all your equipment, all your tools, and then you start," he said. "I was getting eighty dollars a month for six months while I finished the barber college. Well, I worked for ten years in Houston as a barber for two shops."

Carlos found a job with Norris of Houston, a famous local hairdresser. He served the Houston business as a manager in the 1960s, but he had bigger plans.

The entrepreneurial young Carlos Villagomez was looking for the right opportunity to build his own business. As President Kennedy had spurred the growth of NASA, the innovative thinkers developing the space program understood the need for pioneers to create a new space community to support the massive effort. Gene Horton, the NASA public affairs director, invited Carlos to become part of the NASA community.

Horton was part of the original Space Task Group created to manage America's human spaceflight program. He recruited small business owners to support the growing NASA community, forming a vibrant society surrounding the future Manned Space-craft Center.

Carlos recalls Gene's pitch: "He says, 'Carlos, come work at Clear Lake, the new frontier. We need you.'" He checked out the nearby rural area of Webster, which was best known as a speed trap where most experienced truckers knew to slow down as they plowed through. It was remote and unfriendly:

> Just a little old town, you can hear the wind blowing; cows all up and down next door because the trains used to come there from Galveston and they were loaded with wheat, and those [trains] would jerk back and forth and they dropped a lot of wheat [which attracted] deer and the cows.
>
> The people were just farmers. And then they were mixed in with all the engineers that were coming in. Two different groups. And they just didn't get along.

The rural Texas site was chosen largely because of the power and influence of then-Vice President Lyndon Baines Johnson. As a powerful Texas senator, LBJ had sponsored the law creating NASA. (In 1973, after his death, the center was renamed in his honor.)

Seeing the dusty, rural area and the planned NASA site on the cow pasture near Webster, Carlos took an intermediate step: move with the astronauts and early personnel to South Houston, where he found a temporary shop location.

After he set up shop, Carlos met the Mercury astronauts. Alan Shepard was among those early clients. Carlos and Shepard became friends after about three visits, leading to golf and other pursuits—including some solid entrepreneurial advice from Shepard to Carlos on buying land—just as soon as he could afford it —in the areas near the new Space Center.

Most of Carlos's potential clients were housed temporarily in South Houston near the region's first shopping mall—Gulfgate Mall in South Houston. Carlos explained:

> I was working at Gulfgate and there's a place there right across from Gulfgate between here and Houston, and they call it Office City. Somebody had invested and built a bunch of buildings. They're still there. And NASA came along and rented them all. So they were bringing all the people there.
>
> And my shop was right there at Gulfgate across the street. So that's how I met Gus Grissom first and then Buzz Aldrin next. And the third was Alan Shepard.

NASA COMES CALLING

Part of Horton's pitch to Carlos was that the military men at NASA were giving up their flat tops for longer, more contemporary styles, and the area needed a good barber. After Carlos enjoyed success at his temporary first shop in South Houston, Horton pushed him to move to the permanent location near the Manned Spacecraft Center. Horton had scouted a room in a Holiday Inn in Webster, across from the emerging Spacecraft Center.

Carlos visited the hotel. A "hippie hairdresser" had abandoned the room, which had apparently been used for various endeavors.

Carlos eyed the room, with the remnants of an abandoned

dental clinic and dental drills and other devices, which had all somehow been converted for use as a barbershop, complete with a strange dental chair. Carlos explains he had never seen anything like it.

The Holiday Inn required a $200 deposit, presumably for potential damage to the remnants of the dental clinic. He eventually bargained for a month-to-month lease of $240 a month. The hotel told him he would have to leave if it needed the room. He paid the hefty deposit and came to Webster while NASA was still under construction on the cow pasture, knowing he was working with pioneers, answering the late President Kennedy's call to put a man on the moon before the end of the decade.

Carlos figured he'd use the odd dental chair until he saved enough for a real barber's chair. "I could cut hair with that, and I did for two years until I made enough money to buy this chair," he said.

He gestured to the handsome chair in which he was proudly sitting, so many decades later: "Fifty-three years ago, I bought it from Japan. And then another. They're Belmonts. They're the best."

Carlos achieved immediate success in barbering to the NASA crowd:

So I say I am going to give it a whirl. It got good right away. And they were working very, very, very hard, the seven astronauts, because they all wanted to be first up.

Gus Grissom ... liked my flat top. He used to wear a super short flat top. And Shepard ... we hit it off as friends. And of course, he was eight years older than I was, and super smart.

I wasn't used to hanging around with those kind of people ... He was nice to me, and I was nice to him. After meeting two or three times, we became close enough friends that we'd go have a beer and then he would ask me a lot of questions about who's the [golf] pro at a certain [place].

He wanted to know everything, so that he knew what's going on. So now he told me, "I want you to be my eyes and ears."

Shepard and Carlos were entrepreneurial, intelligent men who saw the potential in the area.

Shepard got together with some other people, and they started building South Shore Harbor. And then he took me out one day and said, "Carlos, I'm going to put you on our MUD district—municipal utility district—because I want you to be my eyes and ears in this area in South Shore Harbor."

And he said, "if you have any money, Carlos, buy all the real estate you can around here and here." And he was right.

Carlos said he was glad to help his friend with information. "How can I give Shepard the astronaut credibility, and I'm just his barber? I think one thing I did give him was a lot of information."

Shepard knew what was happening in the community in which he took such interest. Carlos explained he knew Shepard pursued information: "He was well-informed because he was telling me everything that was going on in the area with every politician up and down here."

OUT OF THE MOTEL, INTO THE BOARDING HOUSE

As Carlos built his business at the Holiday Inn, word of his barbering skills and abilities spread within the NASA community. Carlos worked and saved from his venue in the strange little hotel room. A few years later, he needed more air conditioning and more space, and the Holiday Inn had neither. He had to move on.

He turned this forced move to advantage, finding a rambling old boarding house in the town of Webster, the dusty town he'd first scouted years earlier.

When Carlos moved onto his newly purchased property, boarders still occupied many of the fourteen rooms in the aging building. Within a few years, they drifted on, and Carlos replaced the boarders with barbers. He welcomed diverse barbers to set up their unique shops in the various rooms, most decorated in unusual, distinctive styles.

He began buying all the land he could afford around the old boarding house for parking, growth, and real estate investment.

During the sixties and early seventies, Carlos and his barbers became established members of the community including farmers, industry workers, ranch hands, bikers, country folk, and newer service people, engineers, astronauts, and aerospace contractors who had settled locally in new subdivisions nestled near lakes and old bait shops, roadhouses, cow pastures, and newly arriving businesses.

The entrepreneurial Carlos bought a switchboard, which helped in making appointments and added to his profits by selling answering services and private line connections to over a hundred customers. He staffed that old-fashioned switchboard with an operator twenty-four hours a day, seven days a week.

Other entrepreneurs flocked in, too, hoping to capitalize on the emergence of a space city. The main drag was NASA Road 1, a street that would be flooded by subsidence and the lake it crosses, to be rebuilt every so often and ultimately replaced by what you can visit now via a broad fancy ribbon of concrete. A few years later, Carlos further developed the large piece of property surrounding the aging boarding house, which he eventually demolished.

You'll find the property in Webster, Texas, on Old Galveston

Road. That old road has been enlarged over the decades and is now more commonly known as Highway 3.

Nowadays, Carlos rarely cuts hair. Rather, he observes his protégés from his tiny "shotgun-shack" on the property and holds court at a picnic table in the back of the Beer Garden. Yes, you could still shoot through the barbershop front door, and the shotgun pellets would go out through the back door.

A SPEED-TRAP REPUTATION

Back in the old days, the town of Webster clearly needed to outgrow its speed trap reputation. Most locals and truckers knew to slow down as they plowed through.

Good people back then saw the need for reforms, recognized Carlos's potential, and asked him to help.

Carlos spoke to the police chief and mayor to make the little town a better place to live for everyone, but his first efforts met with failure. People with power and town old-timers held to the Wild West cowboy mentality; law enforcement and the fire department were dominated by men who felt they were the law. Such men felt no need for input from the populace. They had no respect for energetic young elected officials or their appointed chiefs who failed to defer to the old guard.

Around the same time, Carlos decided to use a good part of his property to develop the Beer Garden, which enjoyed immediate popularity. Carlos Beer Garden was (and remains) a stone's throw from the Webster Police Department.

Rogue Webster police officers were not pleased, and they worked to make business and life difficult for their neighbor. One evening, heavily armed local police burst into the new Beer Garden, waving pistols and wielding their shotguns as one officer drew his finger across his neck in a slicing gesture.

"Cut the noise!" they yelled.

They menaced customers, threatening them with jail, a terrifying prospect for the NASA engineers, businessmen, hardworking Texas City plant workers, ranch hands, and bikers gathered at the Beer Garden.

What did I do? They were threatening my customers and my bar people! So I snuck out the back and ran to the mayor's house. I knew the mayor would be fair, even though he wasn't that happy when I tried to get a license to sell beer in my place next to a police station. The city council had been divided, three yeses and three nos. The mayor hadn't liked the idea.

But I remembered: the mayor had said that if I operated within the law, a man should be entitled to sell beer there. He had voted yes, and that's how I got my license. He was my best chance.

When I ran over to the mayor's house that night and told him what happened, he was angry! The next morning, he called them into his office and fired those renegades. And the mayor wrote a letter to the chamber of commerce folks telling them they should welcome me.

The mayor's willingness to accept the determined Carlos and his new Beer Garden was not matched by the community.

Back then, I was living on top of the old boarding house. That same night, someone shot up the place, the doors. Man, they shotgunned all over the whole front. My switch-board operator was working the night shift on the first floor. They could have killed him! These people were really hurting business.

The police chief came to talk to Carlos the next day, he recalls, as he was assessing the damage: "He begged me, 'Please don't call the FBI.'"

Carlos considered his situation. He stayed and grew stronger from the roots he had planted in the NASA community:

> I knew they'd fired those cops, so I paid for the damage they caused. No one else would do it. And I repaired the mess they made. I told them I wasn't going away.
>
> That was in the early 1980s; that's when I decided I needed to help this town change. So I ran for city council. The first time, I lost by one vote. But then a councilman resigned, and the mayor appointed me because, well, I lost by one vote. I was the logical choice.
>
> So I was then put on the council. After that, I got reelected and then they made me mayor pro tem.

Carlos learned much serving on the city council and as mayor pro tem. He remained active, also serving as a leader in his church and Rotary Club. He sought wise people to learn how good government can best function, particularly on a city or town level. He put what he learned to use:

> We needed a strong manager form of government, a person with a spirit, experience, and expertise who wouldn't leave after an election cycle. So I strongly suggested to the council and mayor that we adopt a strong manager form. I proposed we amend our documents, and it was first disputed. And we debated it some more. But then, it passed!

BATTLING RENEGADES AND CROOKS

Carlos recounts the arduous search, bitterly resisted by the old-timers, for a truly independent manager. He found a great one, out of state. Carlos then provided much information to the expert manager, who listened and carefully examined the town's books and records.

The new city manager understood from his review how the town of Webster had been operating, renegades and crooks included. He also taught Carlos about methods and processes for economic development. Carlos speaks efficiently and eloquently about the importance and detailed functions of accounting in various city departments. The city manager, he explains, "started to investigate those officials in our fire department who were supposed to be keeping proper records. He learned that two of the trucks had been confiscated by the IRS, and the fire department wasn't even in possession of equipment it claimed."

Carlos recalls what happened next:

I revealed a massive scandal in the fire department to the city council, and since I was also mayor pro tem, we were able to pass a tax for infrastructure and business development which still aids our current city team. It is great for development.

We created the Webster Economic Development Board. I'm very happy about what happened to the city of Webster.

Now in his late eighties, Carlos remains on the Webster Economic Development Board and treasures the businesses and roads, solidly built homes, sidewalks, and exciting new developments, all resisted by the old-timers.

Carlos's leadership in the community is why, on any given day, you might exit the barbershop and step into the Beer Garden and have a burger and fries with the president of the Bay Area Houston Economic Partnership, visiting with Carlos and local church leaders about the Spaceport and a long list of developments in the now thriving region.

BARBER TO THE STARS

In his heyday, Carlos cut the hair of major press figures, movie stars, visiting dignitaries, and many women enjoying boyish styles.

He cut the hair of many female astronauts and friends in the new groups within the astronaut corps; he was glad to see the growing diversity.

He grieved mightily after the Challenger disaster when he learned what had happened to his friends. He recalled with much fondness that not long before the disaster, an exuberant female astronaut on that flight had visited with buddies and borrowed billiard balls from the Beer Garden pool table for a scientific experiment in space. Although they were on board the Challenger, none of Carlos' billard balls were recovered.

Carlos watches John—his trusty barber and scheduler—as the younger man cuts and grooms hair, eyebrows, and the magnificent mustaches and beards of everyday heroes from local homes, hospitals, schools, hotels, farms, businesses, and bustling life around the Johnson Space Center.

Carlos has a history of taking younger barbers under his wing. He and the Bay Area community recently grieved over the untimely death of Jesse Salinas—a master barber in his own right— who worked alongside Carlos, carrying on the traditions. Jesse spent his spare time in humanitarian good works in the Galveston Bay area and dreamed happily of a run for city council, which he was planning under Carlos's guidance. New, younger barbers, including John, now follow in Jesse's footsteps.

The COVID-19 crisis did not stop Carlos and his wife, Rachel, from delivering thousands of tamales, as they do every Christmas. His family and his employees gave away thousands more in the hard winter of 2020 and 2021, and the family—now including great-granddaughters — carry on the traditions.

For more than forty years, he has served his Rotary community and their sister club in Mexico City. On a visit there, he learned of the great needs of a medical clinic built by a donor, who left it unfurnished.

Carlos vowed to outfit the empty clinic building. He came home and visited with friends, including a popular local hospital executive, Raymond Khoury, who welcomed the request. The Methodist St. John's Hospital administrator, who now teaches hospital administration at the University of Houston, had recently upgraded the hospital with brand new equipment, storing the slightly used, still state-of-the-art monitors, beds, sterilizing equipment, wheelchairs, x-ray machines, and so much more.

Khoury provided anything one would find at a modern hospital. The equipment and supplies filled ten huge storage units.

Carlos asked at his Rotary meeting for an eighteen-wheeler and driver to drive the equipment to Mexico, all needed for the following Saturday. That mission was also accomplished, with thanks to a generous hauling company. Carlos arranged for loading, with Rotarians and community friends pitching in. They loaded the eighteen-wheeler to the gills that Saturday, packing every space with additional donations of supplies, including baby formula and diapers. The massive load departed, on time, with its vital contents secured.

An inspired ladies' club held a fundraiser to help, too, honoring Carlos with a special function arranged to buy more supplies, which led to more philanthropic work, as the community recognized the great need for specialized medical care for children with serious cleft palate and burn injuries. Thanks to Carlos's initiative, doctors—specialists including dentists and facial surgeons—stepped forward to volunteer their much-needed services.

Community members opened their homes to the families arriving with children requiring special attention and needing places to recover—sometimes for months before the children and parents could safely travel home. Decades later, these grown children write and visit Carlos as healthy, handsome adults, with children of their own.

Carlos is a great example of how good people make our communities work. They spend their life in good works, charitable efforts, and helping strangers and friends, beyond their worlds of comfortable barber chairs, beer gardens, sage advice, and listening. Carlos shares his wisdom and his bounty. This barber's life of good deeds and problem-solving has benefitted generations of deserving people.

Carlos continued to meet more generations of people saved by his warship during his military service. Recently, when Carlos went into the hospital for surgery, a physician examining him noticed his Navy tattoo. The doctor told Carlos the story of his grandmother's treasured account of survival in her youth, a family unrelated to the folks he met years prior.

The grandmother ensured her descendants preserved her memories of a U.S. Navy ship with a "big gate," the ship that rescued her as she was escaping what was to become known as North Vietnam, Indochina, in the 1950s.

Her grandson wanted to thank every member of the U.S. Navy he encountered. He wanted Carlos to understand that he owes his very existence to the work of the U.S. Navy as he is a descendant of one of the thousands of lives saved, because of the work of a great U.S. naval vessel and its crew.

BUILDING A BETTER BEER BUSINESS

As already described, when Alan Shepard obtained a Coors beer distributorship, with his former test pilot buddy who was married to a member of the Coors family, Shepard turned to Carlos for experience in the world of beer sales in Texas.

Carlos knew from experience that the Coors cans tended to rust, and—frustratingly for everyone physically involved—the Coors delivery folks were required to rotate the older cans at every bar they serviced to promote sales of those cans before the rust appeared. He explained to Shepard the disruptions and problems created by the Coors delivery drivers moving around individual beer cans behind the bar during business hours. The drivers took longer to perform the task than what a good barkeeper could do more efficiently.

Carlos advised Shepard the distributorship should stop the hugely time-wasteful process of rotation, which delayed the drivers' travel and delivery schedules. Trust the bar owners to do their job in handling the inventory, said Carlos. Those Coors delivery men, in rotating inventory, were disturbing the hardworking barkeepers and customers.

Also complaining were the patrons of Maribelle's, a famous Kemah waterfront bar, painted pink and visible from the Kemah Bridge, offering patrons entertainment and games of chance. Maribelle herself asked Carlos to solve the problem. The disturbances—and loss of privacy for customers—were hurting everyone's business.

Shepard asked Carlos to come into the nearby Coors headquarters in Deer Park to explain the problem. As requested, Carlos drove over to the local Coors distributorship to explain the business interruption losses. Who did he find waiting for him? None other than a member of the Coors family, wanting to hear the issue.

After the visitor received the briefing on the disadvantages to customers, and the bars' abilities and desires to handle rotation of their own Coors stock, Carlos says, Coors changed the policy on a broad scale, to the relief of all, including the hardworking ladies at Maribelle's.

Carlos said that, thanks to Shepard and his partner in the Coors distributorship, Maribelle and Carlos took another issue—safety—to the Coors folks. Carlos tells the story that Maribelle had asked him to advocate for a solution to the problems her lady bartenders suffered while opening Coors cans with a push-through, stamped opening, requiring a finger press, resulting in a hole with a ragged edge and sliced fingers for the servers who opened them.

Carlos went to Shepard, identifying both the problem and a

solution: use of a tab to distance fingers from the metal edge. Shepard brought the issue and proposed solution to the attention of the Coors folks. The company ended up adopting a safer tab. Carlos smiles proudly as he recounts the delight of ladies opening beer cans behind his counter and serving at Maribelle's.

THE SPANISH TEACHER

Shepard was fascinated with the Hispanic community and keen to learn the language. He asked Carlos to teach him, starting with phrases he asked Carlos to translate, to get him started. He was an excellent pupil. Shepard embraced every opportunity to meet Carlos's friends and family from Mexico, wanting to practice his Spanish.

Carlos recounted the amazement of his loved ones at Shepard's generosity with his time and at his serious interest in learning and communicating in Spanish and learning Mexican culture. Carlos tells many stories of friends and family gloriously happy to be invited to meet and speak Spanish with the famous astronaut who had walked on the moon:

> They could not believe it. So I called Shepard and he right away would say, oh yeah, come over. You see, I'd take them over and, oh, he loved to speak Spanish to them, and have a lot of fun with the whole family and take a lot of pictures with them.
>
> In fact, when I went to Mexico to visit some of my aunts—way back in the seventies and eighties—every home I went to had those pictures ... everybody, because they took so many pictures.

Carlos says these happy photos of Shepard still dominate the walls of his friends and family in Mexico, in places usually reserved for religious icons or matriarchs and patriarchs.

CARLOS BEER GARDEN: A REFUGE

After his first few astronaut customers, Carlos went on to cut the hair of many more astronauts of that era—household names still. As is clear from the memorabilia on the walls, his popular barbershop (and more popular beer garden) welcomed astronauts of the Mercury, Gemini, Apollo, Shuttle projects, and beyond. To this day, both businesses welcome many customers from NASA.

Carlos is proud his establishment and the surrounding land have always been a place of comfort for friends, including America's first man in space. Shepard found refuge strolling in the back areas and speaking to Carlos's never-ending crew of Hispanic day-laborers, practicing the Spanish he asked Carlos to teach him. Shepard was determined to become fluent. Carlos was impressed.

Among the laborers working with Carlos in the NASA area, Shepard exhibited special patience, posing for photos, making small talk, and joking. When he returned from the moon, he wanted just to visit without being "on," said Carlos. He would take breaks, when possible, just to relax.

Carlos hired day laborers for work on building and maintenance on the land, putting in trees and improvements behind the Barbershop and Beer Garden. They were working high up as Carlos and Shepard were walking outside. Shepard called up to the workers, joking around in Spanish.

Carlos let the workers know who was chatting with them. They couldn't believe it; here they were talking to Alan Shepard! Every man overhead knew they were talking to the man who had recently returned from walking on the moon.

Shepard was calling up to them, in Spanish, with the cadence and accent he'd learned from Carlos, bantering. The moonwalker fluently answered the workers' questions about his lunar explorations, on what it was really like "up there," in Spanish, with perfect timing:

Too hot.

Too cold.

No girls.

No beer, either.

Everyone roared from the sky above the astronaut.

DISGUISING THE FIRST MOONWALKER

Carlos knew Neil Armstrong from the early days, but he wasn't a regular customer, unlike many of the Mercury, Gemini, and Apollo astronauts. But the first man on the moon knew to come to Carlos with a specific issue. Armstrong kept his hair short, parted at his receding hairline when it was not cropped. The famed spaceman's hair was barbered so closely as to require no combing. This quiet, reserved gentleman told Carlos he wanted to travel, but his face was so famous he could not go out publicly and retain his privacy. He could not enjoy the great pleasure of anonymity in traveling and sightseeing. Carlos, as always, listened well and then explained what a huge change one's hair can make in appearance.

Armstrong said he wanted to travel in Mexico. Carlos recommended he grow a mustache and let his hair grow quite long. Armstrong returned with long hair, but he could not grow a good mustache. Carlos bought a luxuriant false mustache, trimmed the astronaut's hair and mustache into a Pancho Villa style, then dyed everything to near-black. Armstrong learned quickly how to glue and groom his new appearance properly.

The world-famous face was unrecognizable. He enjoyed a long and reportedly wonderful private time in his travels through Mexico.

Carlos enjoys friendships with new generations, including the modern astronaut corps. Yet he remains a beloved community member and friend to the astronaut corps and NASA community; he and old friends remain in touch.

BARBERING FOR A NEW GENERATION

In learning more about Carlos, we were honored to meet virtually several times with his dear friend and shuttle astronaut Joe Allen. In the late 1960s, the gifted young physicist received a coveted invitation to join the astronaut corps—via a phone call from Alan Shepard. With much delight, Joe Allen tells the story of reporting to Shepard's secretary at the astronaut office in the Manned Space-craft Center in Houston only to be mistaken for a trespassing schoolboy, who must surely have evaded a field trip to invade a secure location at the Space Center.

The brilliant scientist—who held a Yale doctorate in physics, who would soon receive top honors in his NASA-sponsored mili-tary flight training, and who would progress to fly in space—was ordered to leave Shepard's office immediately. The new astronaut searched his pockets to fish out his orders.

Allen served as CapCom on Apollo and Space Shuttle flights before his own Shuttle missions, and he also served as head of NASA's Congressional Liaison Office in Washington. More recently, he swapped tales with Carlos over more than one Zoom call, as they reminisced about Alan Shepard and the old days.

Allen most famously made aviation and Space Shuttle history by wrestling with and retrieving from space a malfunctioning satel-lite—and returning it to the Space Shuttle Discovery. Allen also wrote a spectacular astronaut book, described in the Sources section of this book.*

THE BARBER AS AN ELDER STATESMAN

The current generation still asks for the stories from the 1960s and 1970s, when Carlos's barbershop was a destination, whether it was the strange Holiday Inn hotel room, his shotgun-shack barbershop

---

* *Entering Space: An Astronaut's Odyssey,* 1984

(still in use next to the metal hulk of the Beer Garden), or the short-lived satellite salon at the famed Warwick Hotel in Houston, where he cut and barbered celebrity and society folks, in the heyday of Apollo moon missions.

Nowadays, Carlos cannot stand on his feet long enough to barber properly, but he's mentoring a new generation of barbers who never tire of the stories of the many clients of the Barbershop and Beer Garden, including Mercury, Gemini, Apollo, Shuttle and contemporary crews, plus movie stars, royalty, farmers, ranch hands, oil field workers, bikers, lawyers, doctors, contractors, scientists, tradespeople, and local families of Webster and Kemah and Clear Lake City, Texas.

Ask Carlos about trimming and styling John Wayne every day for a movie being shot nearby, or barbering the Duke of Windsor after he abdicated from the throne of England. Or cutting the hair of Andre Previn and his actress wife Mia Farrow, who lived in Houston when Previn conducted the Houston Symphony.

Carlos holds court from his barbershop chairs or Beer Garden picnic tables with stories of cutting the hair of early movie idols Rory Calhoun and Johnny Weissmuller, the star of the original Tarzan movies.

Local folks regard Carlos as an elder statesman in his community, a peacemaker across all political lines, and a respected businessman, former city councilman, and mayor pro tem of what was once a tiny, insular rural community.

Carlos was greatly moved by an astronaut friend walking into the barbershop around Christmastime after suffering a great loss. In his grief, the astronaut took Christmas lights and tinsel over to the barbershop—and just started decorating the place. It is a very meaningful memory for Carlos. He also remembers a call to his barbershop from aboard the International Space Station.

In the evaluation of retired NASA test pilot and SR 71 Blackbird pilot Terry Pappas, Carlos is the friendliest man he's ever known.

Carlos will never forget the extraordinary evening at the home of Alan and Louise Shepard, when he cut the astronaut's hair before Apollo 14 isolation began. Carlos most enjoyed drinks and listening under the stars and moon, as Shepard asked his friend to pepper him with questions, explaining what he planned to do "up there," pushing Carlos to ask him more and more questions.

Carlos asked Shepard to "do something" for him up there, maybe write the barber's name in the sand, and received only the famous Shepard grin in reply.

Carlos explained in our documentary interview that the next step in the plan was to schedule a haircut as soon as possible after the lengthy mission.

"'I want you to be available,'" Carlos recalls Shepard telling him. "'When I come back, I'll come straight to your shop.' That's exactly what he said. And what he did!"

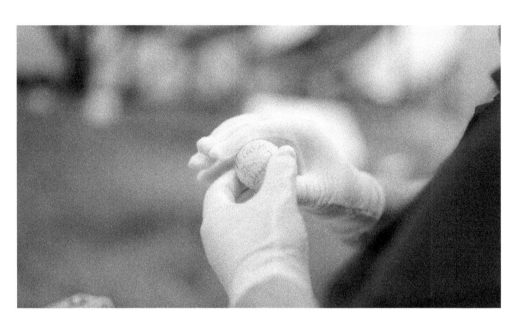

CHAPTER 3
THE GOLF BALL

ON THE PAVEMENT outside Holiday Inn room number 105, Carlos watched Alan Shepard—arguably the most famous man in the world upon his return from the moon mission and out-of-this-world golfing episode—with his big, famous grin and happy wife. The couple walked hand-in-hand toward him.

Alan Shepard and Carlos Villagomez shook hands with a big, firm handshake, their eyes meeting. Carlos recalls his surprise at feeling a hard, round object in his hand. The barber then realized that a smiling Shepard had just palmed a golf ball into his hand. And his friend was still grinning—more widely than the famous Shepard grin.

Shepard provided only a few words and smiled broadly—"lots of teeth," Carlos remembers. His memory of the words is not perfect; he thinks Shepard said something like, "This is for you and your wife," and "Take good care of it."

The location, the supreme happiness on everyone's part, the memory of the initial handshake—and the feel of the handshake surprisingly accented by the hard, palmed ball—Carlos remembers it well.

The hair of the famous astronaut was—as expected—a colossal

mess. Carlos's explosively vivid comment in our recorded interview that Alan Shepard's hair looked "like a firecracker had gone off" brings to mind the Looney Toons cartoons of the day that both family men knew well; Roadrunner and Wile E. Coyote provided no shortage of visuals of the effects of firecrackers on the hair of popular cartoon characters.

Carlos marveled that he was holding a golf ball from the most famous man in the world, who had returned from golfing on the moon, a sporting event on everyone's mind. Immediately, the trio entered the barbershop and got down to business. As soon as Shepard was seated in the barber chair, he said, "Lock the door, Carlos."

Carlos never asked Shepard about the ball or its story. And for the rest of his life, Shepard never raised the matter, either. Carlos took it to a safe deposit box.

Carlos has no photographic proof of the intricate process involving hair gel, hair spray, and the ladies' bubble dryer. Carlos has not one photo of cutting or styling Shepard's hair. He has no photo of Shepard looking anything but perfect.

Did Alan Shepard take three golf balls to the moon, using only the two that remain there? And, if so, did he give his barber a third ball he didn't use?

We gained much insight on this journey from several respected figures, including the space historian and memorabilia expert Robert Pearlman; Shepard's daughter Laura; Apollo moonwalker Charlie Duke; and NASA/military test pilots and astronaut friends of Carlos.

And we also contacted the oldest and highest level of key Apollo era Spacecraft Center administration: George Abbey, also known in the Apollo era as "The Astronaut Maker." At the time of our interview, Abbey, a friend of Carlos since the early days of NASA, was alert, active and spry. He repeatedly demonstrated his famous, prodigious memory as he reminisced. We interviewed him at the Beer Garden just before his ninety-first birthday; he was the last of our interviews for this book. Please see the Acknowledge-ments for more on the late, great George Abbey, who passed away on March 24, 2024.

We filmed our first interview with Pearlman next to the Saturn 5 rocket, which is the most prominent feature of the "George W.S. Abbey Rocket Park" at the Johnson Space center.

NO PROVENANCE

The world watched as the golfing astronaut unquestionably left two golf balls still on the moon. Robert Pearlman made clear his opinion that Shepard brought aboard the number of balls declared on the ship's manifest: two.

Pearlman understands the business of memorabilia well. You'll see his extensive experience and leadership in the acknowledg-ments at the back of this book. His thorough search of the records located the Apollo spaceship's manifest, which listed a declaration of two golf balls. We know where those two balls lie: on the lunar surface. Did Shepard bring a third ball in reserve, in case he twice missed? Did he bring it home to his barber?

Pearlman noted Shepard's admirable discipline and military background, emphasizing his dedication and commitment to duty. And he commented that as a leader, Shepard would set a good example for the astronaut corps he had led. Pearlman emphasized that Shepard—medically cleared to active duty and readying to fly the Apollo 14 Mission as commander and pilot of the lunar lander —would do nothing to endanger the mission, such as bringing excess weight. Was the astronaut entitled to take a third ball, not listed on the manifest? He found no document proving that Shepard received permission to take a third golf ball. Did he receive permission? There's no written proof, no signed "prove-nance." And, Pearlman asks, why not test the ball for moondust?

Pearlman told us evidence that Carlos's ball flew is simply inadequate. But is it? We decided to dig a little deeper.

Pearlman noted when Shepard "said he had a 'white pellet' that people might recognize, implying just one." This interpretation respects Shepard's care in language choices; the astronaut gave himself that flexibility. We agree with Pearlman's observation that Shepard's "intention was that if he actually swung in and connected with it and it went 'miles and miles,' then that would have been the end of it."

Shepard clearly knew he needed a second ball, a "reserve" in case anything went wrong. He produced that second ball before a worldwide audience and connected well with his second attempt.

Although Pearlman doubts the Carlos golf ball flew, he agreed to examine it and validated Shepard's handwritten signature and gifting language "To Carlos."

In his career, Pearlman has examined many golf balls signed by Shepard. He explained the astronaut was often asked to sign the object for which he became best known, and he signed lots of them on and off the golf course. Pearlman validated the ball signed "To Carlos," bears genuine Shepard handwriting and signature.

The memorabilia expert emphasized—and we agree—that Shepard never gave written confirmation that the Carlos ball flew. Shepard signed certificates authenticating other flown items, providing the significance of written provenance.

Pearlman explained:

> The ball itself, it is signed, "To Carlos, Alan Shepard." But if it was actually the flown ball, you'd almost expect that Alan would have written "flown to the moon" on the ball itself. He wrote that inscription on other items that he flew and gifted. So he did recognize the need to document what these items were.
>
> Most of the time, when you see an item that is signed that went to the moon or went to or flew into space, it's the astronaut adding their provenance to the item. The astronaut is able to say, this patch of Velcro looks like every other

patch of Velcro, but I'm going to write on it: "This flew to the moon on this such-and-such a date and I'm going to sign it."

That way you then know that particular item went to the moon—anything that didn't have a serial number or a unique identifier—the pieces that the astronauts generally kept didn't have those. And so, adding an autograph to it or adding a signature in this case to it and maybe an inscription allows that piece to be identified for what it is.

Without written proof of the ball's flight status, Pearlman told us, placing a value on it is difficult.

Not to mark it, not give him any type of provenance to ever establish that—it just raises a lot of questions. The value is a very different difficult question to ask, because it doesn't just appeal to space memorabilia collectors, and it's a piece of sports memorabilia.

It's a piece of Americana. It's a piece of history that spans multiple different interests.

I would not hesitate to say that if it could be established solidly that it was a flown golf ball from Apollo 14, you could see upwards of seven figures being paid at auction, one million dollars or more because of the nature of the sport. [Given] the people that it attracts, a lot of business is done on the golf [course].

With such an extraordinary one-of-a kind piece of space memorabilia, Pearlman explained, he'd expect that the relationship between the giver and recipient would be familial or that of a best friend.

"And he didn't give a golf ball to his wife, as far as we know," Pearlman said. "He didn't give one to his daughters. He didn't give one to his best friend. So extend on that and say, well, then he would give maybe the only other golf ball that he brought back from the moon to his barber."

Pearlman pointed out that we lacked evidence independent of Carlos and his statements of a special relationship with Shepard and a future haircut appointment. He also viewed photographs from the time. He explained that after quarantine was lifted, the press awaited the astronauts. Shepard greeted them, well-groomed and nicely dressed. Carlos has no photographic proof that Shepard looked unruly.

Pearlman said Shepard had access to barbers inside isolation, and Shepard could have used those barbers, avoiding the press as they emerged from quarantine. He was clear: barbers and other service folks—including chefs—stood ready to serve the astronauts.

Regarding Carlos's claim of Shepard's appearance in a flight suit, Pearlman claimed astronauts didn't wear flight suits in quarantine. Rather, he said, they wore civilian clothes.

He said that the press always staked out the area around the quarantine facility to photograph astronauts as soon as they were released. Pearlman searched all available NASA logs for Apollo 14 from space. There is no record of a call discussing a hair appointment. Pearlman also did a thorough search of the official NASA transcript during the mission; he found no call from space to Carlos —nor any request from Shepard in space that NASA call Carlos to remind him to be at the shop.

FALSE MEMORIES?

Pearlman believes this could be a case of false memories, in which people of good faith are caught in a worldwide shared experience and honestly believe they had a role in it. In a conversation after our taped interview, he provided examples of space heroes and people of good faith clearly proven wrong.

Pearlman believes Carlos may be influenced by what he explained is a surprisingly common effect of our shared experiences of profound events. People of good faith can harbor false memories, he explained; he can recount many provably wrong memories of events that just could not have happened. Pearlman spoke to "false memories" by the masses, especially about worldwide events.

Especially with modern television, he notes, certain people wrongly perceive they were actually participants or had a larger role than as passive observers.

Pearlman also noted that the massive publicity surrounding the Apollo 14 launch may indeed have included the barber—which we determined was reported in many local, statewide, and worldwide articles.

He summarized the circumstances and extensive publicity surrounding Shepard and Apollo 14:

> The Apollo astronauts received a lot of attention and, given that this was the first flight after the Apollo 13 accident or problems in space, there was a lot of media attention on the Apollo 14 mission. And so there were a lot of people interviewing Alan Shepard.
>
> Shepard wasn't shy of the camera. He would talk if he was in the right mood for the day. He had an agreement with other Apollo astronauts in the early years to give their story to *Life* magazine. And so it's possible that during one of those interviews, he did mention his barber; seriously, at some point that had to have come out because there was a story on the day of his launch about searching for a barber to give him a new hairstyle from his old buzz cut when he was in the military. And for daily use. And so that story did get out. There was publicity about Carlos being the barber to Alan Shepard.
>
> It just wasn't the story that he had called from space to arrange for an appointment.

"IT PROBABLY WENT TO THE MOON"

Shepard's daughter, Laura, told us she thought her dad probably took three golf balls to the moon; she said he was a "sentimentalist" and honored his three daughters by doing things in threes. Churchley knew the relationship between Carlos and her father, but only through her father. She had never met Carlos, though she certainly knew about him. And her opinion of whether the ball flew? The astronaut's daughter explained:

> I wish I could say yes or no, you know, but my thought is it probably went to the moon. Daddy and Carlos were that good of friends. And knowing Daddy, the sentimentalist that he is ... he had three daughters.
>
> He probably took three golf balls. I have his Volkswagen and the license plate is 007 SSS for his three daughters.

And Laura said in the interview that from her perspective, "I'm sure that he could have done that—giving Carlos the ball."

We asked astronaut Charlie Duke for the perspective of an Apollo moonwalking astronaut about bringing items—including a golf ball—to and from the moon, and gifting based upon relationships. He agreed to the visit, and we were in his living room within forty-eight hours to learn more.

Could—and would—Shepard have brought aboard a third golf ball, not listed in the flight's manifest with the other two, to the moon and even to the surface?

Duke spoke to the point of easily receiving permission to bring small items—such as a golf ball—beyond the two declared by Shepard in the manifest. The Apollo 16 astronaut addressed who could give permission (Deke Slayton) and how easily the astronauts could transfer objects—particularly small objects—that might be taken to the moon's surface.

Duke also spoke about the items brought and the relationships with the people to whom the items were gifted:

He could put it in ... his pilot preference kit and generally you get permission for that; command module bag was sort of unlimited. You could have several pounds' worth. The limit's at the lunar module bag. It was a limit of eight ounces.

But I think once you got airborne, you could take a few of the items in the command module bag and transfer [them] over to the lunar module, so that this actually did go to the moon, and I forget, but I probably had a couple of things that I wanted to get over there that were going to be over eight ounces.

But you know, it maybe added two ounces or so to the lunar module. And I don't think anybody really thought about it. So ... whether he declared two balls or three balls, I don't know. But it probably was all approved for him to do that.

Duke described to us his experiences with his own items and a fellow astronaut's items, particularly for friends and family:

We all had items like that that we took with us for friends and for family and then the rest of the stuff that we took along was items we wanted to have in our possession, that as opportunity arose, you could give them out, to societies and people and as mementos of visits and stuff like that.

I didn't know that Carlos had gotten one.

My PPK had American flags. I grew up in South Carolina, so I had South Carolina flags. I was born in North Carolina. So it was North Carolina flags, [and I had] Texas flags and a lot of friends in Kansas at the time. So I took some Kansas flags, little mini tour flags.

I took a couple ... four of us were in the class of '64 ... test pilot school in the Air Force at Edwards—Stu Roosa, myself, Earl Warden and Hank Hartsfield, [and] three of us got to go to the moon. And so we had we designed this little beta cloth '64 Charlie flag; it looked like a little pin pennant.

Beta cloth flags don't weigh anything. We could stuff fifty in there, and we took some envelopes, the first day covers, things like that, and jewelry for my mom, a pin for my mom and my wife. A couple of things for her—little moonstone ring and little medallions that were about the size of a silver dollar and on every flight, [made] for your crew—on one side, it was the design of your patch with your names. And on the other side, reverse side, it was the date and [who's] who of the flight. So those were taken mostly on a command module because a coin probably weighed in at an ounce or so.

And so you didn't want to overload the lunar module, so it's those kind of things that we took.

He described a touching gift for the congregation of a church.

For Nassau Bay Baptist Church; they did a lot of prayer from there. That's for us. And there was everybody. It was all microfilmed, of course, but everybody in the church, their members, all were on it. Their names were on that.

So when we got back, I gave it back to them and in a nice little plaque. So we had some things like that.*

THE BARBER-ASTRONAUT RELATIONSHIP

We asked Duke his thoughts on whether a barber-astronaut rela-

---

* Interview, April 16, 2022

would be so special as to justify such an extraordinary gift, assuming the ball actually flew, since Pearlman had noted that Shepard didn't give a flown, signed golf ball to anyone in his family.

Pearlman and Duke both emphasized the importance of relationships between an astronaut and the recipient of a flown-in-space gift—and particularly in something as rare as a gift possibly flown to the moon or moon's surface. Would a barber-client relationship suffice? Only if it were very special.

In addition to his church members and other examples, Duke gave examples of the special relationships and bonds between astronauts and certain barbers, particularly those sharing a military bond. Indeed, it was the military bonds that highlighted and best colored the friendly relationships between astronaut and barber and between astronaut and other military veteran friends in civilian life that Charlie Duke recalled—and currently enjoys to this day:

> Most people took personal things. It would mean something to relatives and special friends. [It] wouldn't surprise me at all ... that Alan would have given [his barber] some memento of his flight and the golf ball would have been the number one choice, I think.

At NASA, Duke had no "special barber," he said. In later years, however, he did. The Annapolis graduate and Air Force veteran described a "former Navy enlisted guy," who became his barber after his old one retired:

> And so I found another barber. In this case named Jonathan Ruiz. And he just did a good job. And I liked him. And he left his shop and opened up another, his own shop. And I followed him over because I liked him. So he did my hair really well. So we became friends.
>
> [Friends like that], guys like that at church, they're

former military. Probably my best friend is younger—he's a little bit younger than I am. But he was Navy; worked for ... the Blue Angels ... And we love to hunt together and we just have the same military background, so we meld really well ... And several other military guys are close.

So I can see how that could happen with Al and Carlos —both being in the Navy.

We turned to George Abbey, a longtime friend of Carlos, from the early days. Abbey was an Annapolis graduate, Air Force pilot, helicopter instructor, and engineer. He served NASA in many roles, ultimately becoming director of the Johnson Space Center. He was renown for his prodigious memory and diverse abilities.

His extraordinary career was extensively detailed in Michael Cassutt's biography *The Astronaut Maker: How One Mysterious Engineer Ran Human Spaceflight for a Generation.*

We interviewed Abbey at Carlos Beer Garden on May 25, 2023, and asked him about Carlos's story of a hair appointment confirmed from space by Shepard. We told him the transcripts don't show the call and wanted to know if he knew anything about the "call from space."

"I knew that, that Al wanted to get an appointment with Carlos," he said. "Carlos cut Al's hair all the time. So when Al came out of quarantine, he wanted to get his hair cut by Carlos ... because he hadn't had a haircut all the time he was on the mission."

Abbey confirmed that Shepard and Carlos had a scheduled appointment.

"I'm sure Al confirmed it in quarantine," he said.

We also asked about a "medical channel" we'd heard about, used for certain calls. Mr. Abbey said: "Yes, there was such a channel. It is quite possible it happened on the medical channel."

Asked if Shepard could have brought an extra golf ball in addition to the two the world saw, Abbey replied with a smile: "I think he might have brought more than that, more golf balls."

Abbey specifically referenced a "preference kit" for taking small items to the moon's surface. Like Duke, Shepard had his own preference kit.

"He could bring what he wanted in it," Abbey said. "The Apollo astronauts all had personal preference kits. They came up with the things that they wanted to fly."

Did the Shepard-Carlos relationship warrant the gift of a lunar-flown golf ball?

"Carlos and Shepard were very close," Abbey said. "They did have the relationship. Carlos was more than just a barber. He was a very close friend."

As Ed was checking the recording devices and was about to turn them off, Abbey had kind words to say when Barbara asked about her dad, which prompted a generous response from Abbey (and tears from the interviewer): "Your dad was a tremendous individual at the Center. He played a big role. Particularly after the Apollo fire, when we were developing new materials."

Abbey praised the Crew Systems Team and the "outstanding engineers" in the engineering development directorate. He remembered well the key development, testing, and applications of Fluo-rel, as Ed knows best: Ed's dad conceived it at home with the aid of a junior scientist chemistry kit and a propane torch; both our dads always kept such a torch close at hand in the garage.

"I'D HAVE HAD THAT THIRD BALL IN MY POCKET"

Terry Pappas, the SR-71 pilot, said mission-driven Navy aviators do what it takes to prepare, plan, practice and complete the mission, with leeway allowed to consider the circumstances presented—to such a degree that shoving incredibly expensive planes off the side of a ship into the sea is indeed an option to carry out the mission.

Barbara explained to Pappas—himself an avid golfer in retirement—that the Apollo 14 flight manifest listed only two golf balls, with conditions known to Shepard from pre-mission quarantine practice, where he'd learned the limits of both movement and visibility when trying to hit a golf ball in a bulky space suit. She asked if he thought Shepard might have included a third ball.

With NASA approval of his proposed golf ball educational demonstration—only if the rest of the mission was successful and on time—would Shepard have taken a second reserve ball to the moon's surface?

Navy test pilots' focus on the mission means they feel less pressure to "dot the i's and cross the t's." They concentrate on carrying out the mission, Pappas said.

"It wouldn't surprise me if he'd have had two balls on the manifest and his bringing three," he added.

After the interview, walking across the beer garden parking lot of their cars, Pappas called to Barbara:

"I'd have had that third ball in my pocket," he said. He flashed a boyishly handsome grin and headed off to his golf game.

We followed up with him later, asking about Neil Armstrong's analysis of Deke Slayton and Alan Shepard in the Introduction to *Moon Shot*, where Armstrong praised both Shepard and Slayton as superb test pilots, who understood and emphasized solving problems.

In the book, Armstrong said that as test pilots, Shepard and Slayton pursued "the search for perfection." In that continuous search, pilots "much prefer" to identify problems *before* going aloft.

Pappas understood the "search for perfection" as a Blackbird pilot: "In work situations, say, at Mach 3, at the top of the atmosphere, you only have one opportunity to get things right. It's built into us. Do it right."

Such training and thinking would include the need to "do it right" in planning a golf-on-the-moon mission

"You have eliminated two things we rely on as pilots—both our vision and the ability to feel, in a bulky space suit and visor," he said. "Well, then you make us uncomfortable. With the understanding you're planning for such a situation where you'd be demonstrating your favorite sport, you'd have a third ball in your pocket."

Shepard believed that the golf balls he left near Fra Mauro in 1971 would someday be discovered, reflecting, "Perhaps the youngsters of today will go up and play with them sometime twenty-five or thirty years from now."[*]

Shepard never revealed the brand of the golf balls he took to the moon. Does the brand of the golf balls reflect his fun sense of humor, obvious enjoyment, the purposes of his demonstration—and his talent for showmanship?

The brand name on the Carlos golf ball is MaxFli.

Documentarian Jonathan Richards bought an old MaxFli box from the Apollo 14 time period on eBay. He filmed the opening of the box for a sand trap shot for the documentary (see Richards website, to watch the documentary, with the same title as this book). The newly purchased, sealed box from the time contained three balls, just as golf balls are sold in threes nowadays. They all bear the name MaxFli.

Carlos says he never asked Shepard about the golf ball because it would have been rude, an affront to their friendship, but he always assumed it flew. "The first thing he does for me: gives me a golf ball!" Carlos said.

A CALL FROM SPACE?

He's not clear on details, but Carlos believes he received a call as Shepard was flying home, still in space. The call was a reminder for Carlos to be waiting to do Shepard's hair as soon as quarantine was lifted. Carlos does not remember just how it was conveyed. To this day, he believes he received a reminder from space of the haircut

---

[*]   Interview, February 1, 1981

appointment (which was first requested by Shepard and agreed to the night before Apollo 14 isolation began).

Carlos assumed this message was the source of his massive press coverage. However, the coverage appears to have been spurred by one or more of the local or regional interviews of Carlos —picked up by UPI, and spread across the Atlantic.

Was Carlos confused? Carlos says he did not confuse this Apollo 14 reminder from Shepard with much later calls he received from space during Space Shuttle missions. He particularly remembers one call involving a modern-day astronaut calling and telling him to get outside; they were over California and would soon be seen over Webster, Texas. And, he was told, he'd also be able to see the landing if he could get back inside in time and turn on his TV to watch the same Shuttle landing in Florida.

Carlos understands that such modern direct communications were certainly different for the Apollo missions in the late sixties and early seventies.

But he remembers receiving notice as the astronauts were headed home, and a separate one as quarantine ended, to ensure he would be at his shop and waiting for Shepard.

Even if Shepard had access to barbers while in quarantine, Carlos knows Shepard would not use them.

We think it unlikely Shepard would have picked up the golf ball in question during quarantine to autograph for Carlos, and also unlikely that the Shepards would obtain a golf ball and sign it on the brief trip across the street to the barbershop.

Carlos believes the circumstances of the gift are proof enough that the golf ball flew. He never asked his friend for proof, let alone written proof. After the gifting, the two friends never spoke of the golf ball for the rest of Alan Shepard's life.

Did Alan Shepard let another barber cut his hair to prepare him for the world stage after isolation? We admirers of Carlos cannot imagine Shepard entrusting his hair to anyone else, accepting Pearlman's information that there were fine barbers available during post mission quarantine.

Carlos treasures his memories of Alan and Louise Shepard approaching him that day in 1971. Here's what he told us:

> I didn't ask him for the golf ball. He just gave it to me, you know, and I just knew I had something good. I figured it out. But it wasn't until everything settled down that I looked into it and ... I never asked him, "Did this golf ball go down with you to the surface of the moon or did it just stay on Earth—did it just stay in isolation?" ...
>
> I just assume that it had gone with him. I mean, why not? The guy gave me the ball as soon as he came out of the thing.
>
> You know, the most important thing, what everybody's talking about is hitting the golf ball. And it went miles and miles. And then the first thing he does for me: gives me a golf ball!*

While Carlos wondered about the ball's value, he did not ask about it or request "written provenance." In all the years and adventures that followed, the two men never talked about the ball after the haircut post-Apollo 14 mission. Carlos never sought verbal proof, and Shepard never offered it.

After the few words Shepard offered with the gift on his return to Carlos's barbershop, the two men never discussed the ball. Carlos ended our recorded interview this way:

"And so there it is, you know, and I need a drink."

---

* Interview, April 14, 2022

CHAPTER 4
SPACE MEMORABILIA

WE INHABITANTS OF EARTH—LONG-FASCINATED with our skies—value mementos of our interests and travels. The earlier NASA astronauts enjoyed much greater freedom than their more modern counterparts in keeping artifacts and souvenirs of their missions. In practices later recognized by federal law, NASA allowed Mercury, Gemini, Apollo, and Apollo-Soyuz astronauts to carry and bring home mission souvenirs as their own property.

Apollo astronauts kept flown-to-the-moon personal items as well as "artifacts"—including former government property the astronauts brought home from their missions, with NASA's permission.

Federal law cemented this concept, in good part to avoid chaos in the trade of flown memorabilia, and particularly to protect astronauts' practices of laudable donations of such items for worthy charitable causes. The U.S. government recognized the NASA custom and practice by legally protecting those early astronauts and their flown items.

The big exception: moon rocks, lunar soil, and dust. Such items are, by law, federal government property.

We should not forget that the early astronauts tie us today to older eras of aviation history.

The late Michael Collins was a test pilot, Gemini 10 astronaut, and pilot of the Apollo 11 command module Columbia, orbiting the moon during the first lunar landing and exploration. He enjoyed a friendship with Charles Lindbergh, whose 1927 solo transatlantic flight propelled the young aviator into history.

Lindbergh wrote a moving foreword to Collins's memoir *Carrying the Fire,* a book regarded as one of the finest pieces of astronaut writing. Lindbergh wrote of meeting Robert Goddard, the "father of space flight," in 1929 on the porch of Goddard's home, as Goddard spoke of liquid rockets and space exploration.

In his foreword to Collins's book, Lindbergh told of his own profound experiences flying the Atlantic, without sleep for two days and two nights, comparing this with an "awareness" he thought Collins must have felt as he orbited the moon alone, while Armstrong and Aldrin explored the lunar surface.

Collins concluded his book by describing his changing perceptions of the world, as he treasured a letter signed by Lindbergh and reflecting on some of these deep shared feelings. After splashdown, Collins read the Lindbergh letter while in quarantine in the Mobile Quarantine Facility. The astronaut described Lindbergh's letter as "a graduation present."

Collins served the government in many ways, including as director of the National Air and Space Museum and as undersecretary of the Smithsonian Institution. He was a brilliant writer and a thoughtful, kind person. He had a clear interest in preserving space artifacts and memorabilia and using them for education through his Smithsonian service. Collins was kind to Barbara's parents, Matt and Eunice Radnofsky, and stayed in touch with Eunice after Matt died.

TO THE MOON AND BACK

Charlie Duke discussed his Apollo 16 work as lunar module pilot and his lunar explorations with Commander John Young. In our documentary interview, Duke spoke movingly of relationships, including his father's life span covering different historical phases of aviation. Duke's father lived from the era of the Wright Brothers forward, watching his son exploring the moon and safely returning to Earth. Duke made a variety of memorabilia requests, and Deke Slayton approved all of them:

> I took some South Carolina special stuff and a few things for my dad to remember .... He was a big fan of the Wright Brothers (my dad was born in 1907—so, a few years after their famous first flight). And he went through all of the early eras, you know, Lindbergh and flying the Atlantic and 747s and ... and there's kids flying on the moon! He just couldn't believe it.
>
> But he lived to see it and was really thrilled, and so I had a few things for him on board and the kids took a picture of my family that was taken by NASA photography. Ludie Benjamin worked in a photo shop with a little snapshot. And I got Deke [Slayton] and he gave me permission to take it and leave it on the moon. It was the twenty-fifth anniversary of the Air Force, and I was the only Air Force guy going to get to go to the moon that year.
>
> And so they sent me some medals and a little miniature flag, an Air Force flag. And I took those for them and left one medal on the moon, and the other I brought back and gave to the Air Force. Last I looked ... it was on display at the Air Force Museum at Wright-Patterson Air Force Base.[*]

---

[*]   Interview, April 16, 2022

THE SIGNIFICANCE OF SPACE MEMORABILIA

People around the world love to collect space memorabilia—almost anything with a surface to brand, from old souvenirs and postcards to unique, one-of-a-kind flown items.

There's no shortage of personal memorabilia within the NASA community; it seems every family has space-related items: papers, notices, patches printed on special "beta" cloth (developed by our dads and their colleague Frederick S. Dawn for its safe, fireproof, and lightweight properties), autographed photos, books, and other astronaut artifacts. Robert Pearlman explained the beauty and universal appeal of space memorabilia:

> In the history of mankind, less than six hundred people have flown in space, and yet millions of people have dreamt of going there. It's natural because you look up in the night sky and you either see the moon or you see a star.
>
> And it's an unreachable but tantalizing destination. It makes you wonder. It makes you dream. And then when you see other people flying and maybe they're like you, maybe they went to the same school as you. Maybe they grew up in your same hometown, or you're from the same home country. Or maybe they share a similar hobby. You suddenly relate to them and you could see yourself in their adventures. *

Pearlman described his experience with the universal sense of awe in handling a flown object, thanks to his many years of observing the wide appeal of space souvenirs:

> Well, for all of us who couldn't go or haven't gone yet, there are the little bits and pieces, the nuts, and bolts that don't

---

* Interview, April 11, 2022

make it into a museum, that then give you the chance to live vicariously through those adventures.

We will say, I didn't go to space, but I'm touching something that did. And I think that's the universal appeal, even if you don't collect, even if you're not a collector.

I've never met anyone to whom I hand a piece that flew in space who doesn't get a sense of awe like "I'm holding something that went to the moon" or "I'm holding something that flew aboard the International Space Station for a year." That's the appeal to space memorabilia and collectibles.

THE HIERARCHY OF SPACE MEMORABILIA

Memorabilia are similar to autographs in that they can be, as Pearlman describes, a "record of a meeting." As an example, he cited his own autograph from Buzz Aldrin, of Apollo 11, the first lunar landing:

Autographs offer a way for us, in its purest form, to have a souvenir of a meeting. So, it's less the signature or the little drop of ink, but the memory of being able to stand right across from someone you've viewed as a hero all your life, or that you've admired, or that you just audit their adventure .... At its purest form, what an autograph is: it's a record of a meeting between two people. And I think that's why when you have an autograph that's made out to you, especially if it's one that you didn't ask for and you were surprised with, it carries that much more meaning.

I have a lot of autographs, but I have one that hangs on my wall. I used to work with Buzz Aldrin, and one day, he surprised me. He handed me a signed photo. Now, I had handled thousands of his autographs, but he handed me a signed photo. He said, "Hey, I wanted to give this to you.

Pearlman described a detailed hierarchy of space memorabilia, beginning with accessible items intended as souvenirs:

> There [are] the items that were created to be collected that you bought in a gift shop, that you bought in a store or ordered or that were limited edition, and those pieces, you can define their value either by their metal content or by how similar items have sold, or if they are items that have an intrinsic value to themselves.

He explored many types of space memorabilia, particularly items not originally intended as memorabilia or as collectibles, including "the artifacts, the pure artifacts from space flight that were carried into space, that are not meant to be—that were never meant to be—collectibles .... There are bolts, simple bolts that cost millions of dollars to create."

The hierarchy is complex, and the slightest detail can have a big impact on an artifact's value. Pearlman explains:

> There's a hierarchy. Did they go into space? How long were they in space? Were they in space for fifteen minutes on a suborbital flight, or were they in space on an orbital flight and spent a year in space?
>
> When did they fly? Did they fly on the very first missions with Mercury back in the early 1960s? Were they on board Vostok with the cosmonauts? Did they go on Apollo to the moon? And that seems to be something that's flown to the moon. That is the peak of the hierarchy. And then we get into the nitty gritty.
>
> If something that went to the moon is a Holy Grail, then there's a hierarchy within that level, too, which is, did it go into lunar orbit? So you just circled the moon for a while and then came back to Earth, [or did you] land on the moon?

So then within the lunar lander and the lunar module, what went down to the moon, but stayed in the lunar module, never came out, and went back up and came back to Earth, and what [was] taken out on the surface. Did the astronaut have it in his pocket?

Was it a piece of his equipment? And so that's the next level. And then on top of that is, was it stained by moon dust? Now, it's hard to take anything out onto the surface that didn't get stained by moon dust. But if it's something they had that was exterior exposed, then it was most certainly covered in moon dust at some point.

And then, more to the point, you could probably find a photograph of it in use. And you add all those up, then you balance that with the fact that most of those items that went to the moon (there weren't a lot) ended up in museums. You only have a handful of items left over that it's possible for a private person to own.

THE COURTS WEIGH IN

A Texas lawsuit, upheld on appeal to the highest Texas court, resulted in a written decision that honors both the educational benefits of space memorabilia and the charitable work of Laura Shepard Churchley and Charlie Duke.

The law case *Bean v. Bean* emphasizes the large scope of astronaut-flown items, and the important societal interest those items serve beyond continuing interest and excitement in aviation. Federal law protects the early astronauts' rights to their flown objects, many of which are used to support the Astronaut Scholarship Foundation, a nonprofit that provides scholarships to college students in science, technology, engineering, and mathematics. *Bean v. Bean* provides a significant history of NASA customs that were enshrined in federal law, in determining early astronauts' rights—including those of Apollo astronauts—to "personal possessions" flown to the moon.

The Bean case involves a hammer—used on the moon, intended for discard, taken home, and kept after the Apollo 12 mission—by the late astronaut Alan Bean. The case clarifies that earlier "unwritten" rules regarding "personal" possessions could include such discarded government property. The legal decision respects, follows, explains, and provides the history of a special law passed to make sure the Apollo and other early astronauts could keep what they said was theirs.

Duke had testified to Congress in passing the federal legislation protecting astronauts' rights to claim flown items as their own property. Many of these objects funded worthy causes, including scholarships to deserving young people interested in science and mathematics. The Texas *Bean v. Bean* appeals court decision emphasizes the importance of astronaut-donated space memorabilia to charities such as the Astronaut Scholarship Fund. The case includes examples of items much heavier than an extra golf ball that were taken to the moon and brought back as personal souvenirs, a NASA-permitted practice, including the massive hammer in the Bean case.

The Bean case makes clear that the early astronauts of Mercury, Apollo, and the Apollo-Soyuz Test Project were protected both by custom and, later, by a specific law passed to protect the astronauts' ownership rights.

The court was also clear: an Apollo astronaut's property included many items brought home as memorabilia—even though an item might have been a big, heavy geological hammer scheduled to be left on the moon's surface.

On October 5, 2012, Congress passed the "Artifacts of Astronauts Act," successfully protecting the practice and custom of NASA as it applied to the early space missions, confirming "full ownership rights for certain US astronauts to artifacts from the astronauts' space mission." Such artifacts and memorabilia have been used to beneficial effect, including sales for the support of the great educational works of the Astronaut Scholarship Foundation.

Duke's congressional testimony was an important force behind passing this federal law. The Texas case quotes his congressional testimony. He spoke with much authority about the practices during the times of Apollo: as a witness to the old customs, as an Apollo living legend, as a moonwalker, and as a witness to astronauts bringing home objects from their journeys.

The lawmakers used careful wording, which emphasized the law did not grant "new" ownership rights to the older astronauts; rather, the law "confirmed" what the older generation of astronauts had understood for decades: "that they owned and had clear title to the artifacts." (See Bean vs. Bean, detailed in the Sources section of this book.).

THE WIFE, THE ASTRONAUT, AND THE FIRST LADY

We've heard many tales of space memorabilia, but one of the best comes from a barbershop customer—and it's not about the golf ball. It's about a true and loving partnership and fulfillment of a mission —all enabled by a little gold charm one could buy in a souvenir shop.

In our NASA family, understanding of space-related memorabilia has always been grounded in respect and remembering, learning, and love. More than once, I have had the honor of being in Carlos's shop as the late master barber Jesse cut the hair of Naval Academy graduate, combat pilot, test pilot, and astronaut Mike Coats. Coats, a Navy captain, flew 315 combat missions in Southeast Asia. In addition to his service as a pilot's pilot and astronaut, he served as acting chief of the astronaut office. He returned to NASA after service to Lockheed for "the best job in the world": Director of the Johnson Space Center.

At the beginning of his second space mission, with all gear and a few personal items properly packed and stowed, Commander

Coats consulted his wife, Diane, about designating "something" from the crew to be flown for U.S. President George Herbert Walker Bush, the youngest naval aviator in World War II—a huge fan of the space program.

Diane took a week to think things through. She and her husband were aware the president had received numerous special flown items when he served as vice president. She strongly suggested her husband find something small and lovely among the carefully packed personal gear (each member was allotted space for twenty small personal items) as a crew gift to the first lady, not the president.

She pressed her husband to find a woman's gift. It seemed, to the thoughtful Diane, a superb idea. Why for the first lady? "Because if you tell the president you're bringing a surprise gift for Mrs. Bush back from space, we'll all be invited to the White House," was the wise wife's reasoning.

So Commander Coats surveyed the crew; one had a small, gold Space Shuttle charm packed aboard in his personal gear. The astronauts all split the cost of the gold charm and agreed it would be perfect for Barbara Bush. As Diane presciently planned, it came to pass. President Bush called the crew in orbit, as they hoped he might.

The ordinarily reserved Diane was quite vocal and determined that her dream would come to life. She had coached her husband and was now together with the other astronauts' wives, listening intently to the ground link of the presidential call to the crew in orbit. "The wives are watching," Coats tells the story, as he recalls waiting and waiting to insert his casual remark. "And my wife is waiting, and thinking 'Oh no, it's too late, they're signing off, he's not going to ... .'"

Then, as the president was saying his goodbye, Mike said they were returning from space with a little gift packed for the first lady.

"That's fantastic!" the president exclaimed. He insisted the crew come to deliver it in person. "Come here with your wives as soon as this one's over."

After missions, the usual practice was quite clear: quarantine post mission. Not this time. After this mission, the crew and their wives were in Washington within days of landing. They received a separate phone call requesting they come early so Mrs. Bush could give a most memorable, personal White House tour.

In the Lincoln Bedroom, Mrs. Bush went to the window and called out to ask her guests over. White House tourists stood below. She tapped on the window. The crowd went happily wild, and the first lady enthusiastically waved greetings back. "I love to do that," Mrs. Bush said, beaming with happiness.

Following tea with the First Lady, and after presentation of the well-received gold Shuttle charm, the president hosted everyone in the Oval Office. Quite the trip.

Diane quietly left behind a second gift in the Bush White House. She had prepared an outsized dog biscuit in the shape of a Space Shuttle, packed in a shoe box. The biscuit survived frequent openings for curious examination at multiple White House security checkpoints. The still-intact biscuit-in-a-shoebox was left in the dog's room, which the group visited during this memorable tour.

Soon after what would certainly seem the trip of a lifetime, Mike got another call from the White House. Would the crew and wives please return for a state dinner? Invitations accepted.

Diane did not own a ballgown, but they found an appropriate one in the city, albeit three sizes too large. In the little time remaining, the gown was taken in and the crew and their wives were whisked back to the White House.

The Cinderella story continues, as Mike recalls his striking wife outshining the movie stars. You can picture a war hero in full dress uniform and his extraordinary wife standing in the receiving line at a White House state dinner in between Audrey Hepburn and Bob Hope. Next in line is the prime minister of Israel, who is standing next to the president.

Mike was thinking he won't recognize me as the president's gaze wandered over the line. Oh, yes indeed, the President of the United States remembered the thoughtful Diane. Mike recalls the president as he spotted them and called out, "Mike and Diane!"

The president, quite tall, was calling out over heads, and then turned to explain to the Israeli prime minister the story of the astronauts' visit with particular, fond memories of the Shuttle dog biscuit. Mike Coats is convinced the state dinner invitation would not have occurred but for the first family's delight in Diane's gift for the first dog.

As Mike reminisces about his late wife, we learn there is even more to this story about the love of his life, and what her thoughtfulness brought about. Couples were split for the seated dinner. Audrey Hepburn, Dolores Hope, and Mike sat at the president's table. The president wanted to talk about space travel. So he leaned over Audrey Hepburn for more personal conversation with Mike.

"When are you going up again, Mike?"

"I promised my wife this was it. You see, after the Challenger ... we lost a lot of friends on Challenger."

The president—and former World War II naval pilot—pressed on.

"Do you want to go?"

Coats said yes.

The president replied, "Let me see what I can do." President Bush got up from the table and crossed the ballroom. Coats observed him seeking out Diane. She was listening intently to the president.

"She was frowning at me," the astronaut recalls. But then, he says, there was more talk, and he thought he saw her smiling across the ballroom. The president returned to his table, leaned over to Coats, and said, "I think it's a go."

"Mr. President, what did you tell my wife? What could you have said to convince her, she's ... ? "

The president replied, "That's between her and me."

Throughout dinner, president Bush returned often to the topic of space travel; it was such a treat to have the crews visit, he said, lamenting, "What a shame they all don't come here after each mission." The president asked, "Why can't every crew get to come here?"

"Well, sir ... you're the president," Mike replied.

After the extraordinary meal, and still in the White House, Commander Coats took his wife aside and asked her how it was possible for the president to persuade her to change her mind.

She replied only, "We are invited back afterwards."

Diane's generosity in recognizing the importance of her husband's unfulfilled mission, combined with her thoughtfulness, empathy, and strategic thinking, did more than bring to reality the dreams of a White House visit.

Diane Coats' ideas, strategies, and generosity produced a cascade of wonderful results, which required deep listening and thought by all the participants to this mediation, including the president in the middle of a state dinner.

The original agreement between the commander and his wife was based on his deference to her great desire that he'd never risk being lost on another mission.

Diane's desire to shelter her husband from potential harm was overcome with a gesture from a respected figure who reminded her of Commander Coats's need and desire to complete his mission service.

For the rest of President George H.W. Bush's term, every crew member and spouse received a White House post-mission invitation. Diane Coats was the quiet force who instituted a broader tradition most astronauts—and the country—had missed. The commander reminisced that Mrs. Bush wore the treasured, handsome gold Space Shuttle charm on a necklace in an official portrait, paired with her trademark pearls.

Mike Coats left the astronaut corps not long after commanding his third mission into space—and his third trip to the White House.

He tells the story of playing horseshoes with the president in the Rose Garden. Following a distinguished career at Lockheed, he was asked to return to NASA and served the Johnson Space Center well as director. To this day, he still describes that job as "the best in the world."

There's loving proof, then, that space-related memorabilia—a little flown charm and a home-baked, Space-Shuttle-shaped dog biscuit long ago consumed by a presidential dog—changed the lives of the modest astronaut and his wife.

The Coatses, together with the President of the United States, restored a tradition of astronaut visits to the White House, and brought Mike back into space for his third and final mission.

In the barbershop, it's the best love story of all.

A TRIBUTE FROM A COSMONAUT, ASTRONAUT WIDOWS, AND NEIL ARMSTRONG

Memorabilia can be unique and tailored to the recipient. When Deke Slayton and fellow astronauts visited the Russian cosmonauts as part of the Apollo-Soyuz Test Project preparation, they traded gifts, including flags and commemorative plaques. A gifted artist and Soviet/Russian cosmonaut, Alexei Leonov gave the astronauts sketches he had made of them during training.

Deke Slayton wrote about a most meaningful piece of memorabilia, after he experienced the loss of his friends and fellow astronauts in the Apollo 1 fire.

After the fire, the widows of the three astronauts gave Deke Slayton a particular gift, intended as a gift to be flown in the first Apollo mission, to honor Slayton: a unique "Astronaut" pin commissioned by the three astronauts killed in the horrible capsule fire. They intended the pin as a most meaningful gift for their leader: for the medically grounded Slayton, so he would have his astronaut's wings. He wrote about how moved he was by this gift from the grieving widows. They sought to console their husbands' great friend and leader.

On Apollo 11, Neil Armstrong then carried that same diamond-studded pin to the moon and back for Slayton.

SPACE ARTIFACTS ENCOURAGE SPACE EXPLORATION

Special artifacts extend the reach of earlier NASA efforts; artifacts inspire and educate our future explorers. Mike Coats provides a recent example. He escorted Susie Young, the widow of Apollo 16 Commander John Young, to an unveiling of moon rocks as part of an exhibit at the Georgia Institute of Technology Library. The ceremony and effective exhibits honored Commander Young, the university's most famous alum, and highlighted the strong support of the Georgia Tech president and faculty for the work of NASA.

John Young served in the Navy during the Korean War and as a Navy test pilot selected for the second group of America's astronauts. He flew on Gemini 3, commanded Gemini 10, piloted the command module for Apollo 10 and commanded Apollo 16, exploring the lunar surface with Duke. He commanded two Shuttle missions, including the first launch (STS-1).

Young was highly regarded for his engineering skills, and particularly famous in NASA circles for his detailed memos on technical matters, greatly valued by NASA leaders such as George Abbey. In his more than four decades of service, Young also served as chief astronaut.

Georgia Tech's detailed, educational, handsomely presented exhibits, with superb explanations and moon rock displays, impressed Coats. He explained that the moon rocks are important, tangible items inspiring future scientists and explorers. The Georgia Tech library exhibit represented "a great way to capture Georgia Tech students' attention," he said. "The students we met were impressive, enthusiastic, and smart. That's the kind of people we need in the space program. It was a good memory for Susie

Young and me to be greeted by such an enthusiastic response. The government donated the moon rocks in John's name, and the University endowed a chair in the name of John Young."

Coats said he was "proud of the fact that moon rocks have been given to universities [because] they inspire and encourage students to get into the engineering and other sciences."

Toward the end of the Apollo program, the disclosure of a controversy involving back-dating souvenir postal covers carried aboard Apollo 15 led to greater regulation concerning documentation and carrying of space memorabilia, including the listing of items the astronauts sought to carry.

After that tense episode and tighter controls, former director Coats explained, NASA would be less likely to see surprises. One might see items listed as "to be determined"; but overall, the expectations for declared items were more tightly controlled as the Apollo and Apollo-Soyuz test projects wound down. The government tightened its requirements and enforcement for listing the items an astronaut wished to bring into space, seeking more details and specifics.

So, the former JSC director noted, there "weren't going to be any real surprises after that."

CHAPTER 5
DID THE CARLOS GOLF BALL FLY?

IN FEBRUARY 1971, when Alan Shepard stood on the moon and told the world, "I have a little white pellet that's familiar to millions of Americans," the astronaut changed his place in world history.

How many people watching around the world—other than alert flight controllers in Mission Control and Edgar Mitchell watching on the moon—likely caught all his words, let alone nuances, in real time?

Shepard, Slayton, and their journalist co-authors explain the language and the inflections behind Alan Shepard's words during his out-of-this-world experience in the *Moon Shot* memoir.

Shepard whiffed the first ball, after explaining he was trying a "sand trap shot." He explained: "I got more dirt than ball."

The calm golfer seamlessly produced a second ball and—as with the first—dropped the second ball with no ability to see where it lay. It would be another blind shot. He swung again, still limited to an awkward one-handed swing, and was happy to see the ball "sailing" away, as he explains in *Moon Shot*.

Shepard wrote he murmured, "Beautiful." Then he raised his voice, speaking loudly. In a brilliant act of showmanship, he ensured the world would long remember precisely what the golf-playing astronaut exclaimed: "There it goes! Miles and miles and miles!"

MILES AND MILES? NOT QUITE

Just how far did each of two golf balls Alan Shepard left on the moon's surface really travel?

Robert Pearlman explained that we now know—by use of geometry and photographs taken from the lunar module as it was taking off from the moon—how far the two golf balls actually travelled: "The first one that he whiffed went about twenty-four yards and rolled into a nearby crater," he said. "And then the second one went about sixty yards and came to rest. So not 'miles and miles and miles,' but still a good enough demonstration."

Did Shepard take a third ball to the moon's surface and bring it home? And if so, did he give it to his barber? Let's look again at Carlos's great sea tale and the sincere, well-researched response by a respected historian and memorabilia expert. We will answer challenging questions, including:

(1) Why would Shepard add a third ball when two were listed on the manifest?

(2) How would Shepard add a third golf ball to the mission?

(3) Who would be a logical recipient of such a remarkable gift? Did Shepard and Carlos the barber have the kind of relationship to warrant it?

In examining these questions, we accept Pearlman as the memorabilia expert and his best argument: the Carlos golf ball lacks any "written provenance."

Shepard autographed the ball, but he never wrote or signed anything that might prove that the Carlos golf ball flew to the moon. Therefore, we have no written proof from the one person who knew whether the golf ball was in his pocket on the moon, or whether the golf ball flew to the moon. So we also must answer a fourth question:

(4) What do we have, without written proof of flight?

1. WHY ADD A THIRD BALL IF TWO WERE LISTED ON THE MANIFEST?

Why take a third ball? We believe Shepard came to understand, after practice during isolation, that another ball in reserve—the third ball—would ensure a successful experiment and a great show.

Was one backup ball sufficient for the test pilot who taught pilots to never be caught back on your heels? Shepard—ever the smart showman—first implied to the world he carried only one golf ball until he needed that second ball.

Once he successfully achieved the hit on the second ball, and triumphantly announced a "miles and miles" success, there was no need to inform the world he carried yet another ball in reserve. Rather, the world could assume the golfing demonstration had proceeded perfectly, as must have been planned, with two balls. Shepard, never to be caught flat-footed, appeared to proceed with a perfectly executed mission to golf on the moon, right down to knowing how many balls he would need.

Shepard constantly gathered information. During pre-mission quarantine, he gained a sense of the difficulties (in friendly land, in earth gravity) when he practiced in the bulky space suit and helmet. Even on Earth, Shepard could not see the ball where it lay, given the confines of the helmet, nor could he put his hands close enough together to perform a two-handed swing.

Shepard reduced the risk of failure by bringing a second ball. We agree with Pearlman: Shepard had no intention of revealing the second ball, unless he had to use it. With the additional experience, did the seasoned test pilot decide to add a third ball? We believe the answer is yes. A third ball, given its weight, posed no risk to the Apollo 14 mission, and it increased the probability of success of the approved demonstration of lunar golf.

Astronaut Charlie Duke, former JSC Director George Abbey,

and published accounts of flown heavy objects brought home by the astronauts prove the ease with which an Apollo astronaut could bring aboard an item such as a golf ball. Indeed, Alan Bean took aboard a geological hammer intended for discard on the moon, and brought it home. He kept the hammer as his "artifact." That item added weight far greater than a golf ball for a lift-off from the moon's surface.

Abbey and Duke see no reason Deke Slayton would refuse a request from Shepard to bring a third golf ball, additional to the two listed on the manifest. Duke made clear the ease with which permission for such an object would have been granted, explaining the things he also carried. If permission was required, Slayton would have granted it, Duke said. Abbey also mentioned the astronaut's preference kit to bring such small objects.

Surely Shepard assessed the situation after his golf practice during quarantine in the bulky space suit and helmet. Would he have considered his firsthand experience of no visibility and a one-handed swing—and the benefit of an extra ball in reserve?

Neil Armstrong's explanations in *Moon Shot* emphasizing solving problems before the pilot leaves the ground dovetailed clearly with those of Terry Pappas, who explained in more detail how test pilots think. In work situations, say at Mach 3, at the top of the atmosphere, Terry said, "You only have one opportunity to get things right. It's built into us. Do it right."

That's how Blackbird and NASA test pilot Pappas explained why an experienced test pilot would feel the need to "get it right" in planning a golf-on-the-moon mission where "you have eliminated two things we rely on as pilots—both our vision and the ability to feel, in a bulky space suit and visor. Well, then you make us uncomfortable."

So, with the understanding that you're planning for such a situation, he said, "you have a third ball in your pocket." We believe Shepard clearly realized he needed three golf balls, understanding the difficulty of the demonstration—a singular opportunity to both teach and honor the sport he loved.

Given the ease with which a third small object could be added, first clearing it with Deke Slayton and in the straightforward way described by Duke, a skilled test pilot like Shepard would (quietly, perhaps) bring an extra backup.

Before he practiced in the spacesuit and visor, Shepard recognized a need for "dual redundancy" (two balls per the manifest).

After he practiced, Shepard realized he needed three golf balls. He finally had important, additional information on how difficult this NASA-approved, one-handed, blind, golfing-in-a-bulky-space suit gravity demonstration was going to be. Shepard believed deeply in the importance of training and repetitive practice to get as close to perfection as possible. We know circumstances had deprived him of any practice in lunar golf under lunar conditions. The partial practice had made him realize the grave limitations imposed by the bulky space suit, the one-handed swing, the inability to bend over, and the fact he could not see the ball once he dropped it. We know he practiced the swing in a space suit on Earth—but not on an alien surface with extraordinarily different conditions.

For details on NASA's culture of triple redundancy—"a system that operates satisfactorily after two failures" see H. W. Jones's article "NASA Should Not Use the Traditional One- or Two-Fault Tolerance Rules to Design for Reliability." The Sources Chapter provides many more details on triple redundancy and two-fault and three-fault tolerance, including comments from Shepard and other test pilots.

The Apollo 14 crew accomplished their lunar work goals so successfully that the ground bosses authorized Shepard to conduct his dream of a golfing demonstration. He fulfilled his self-imposed obligation to display the game he loved. At the pinnacle of his career, he brought to the task a showman's talent, a scientist's devotion to testing theories, and a Navy pilot's compulsion to "do it right."

The obvious solution, for a test pilot with a specific set of missions that included a world-famous experiment involving the sport he loved: bring an extra reserve ball. If he didn't use that last reserve, so be it.

Faced with an absence of repeated practice and no ability to train for lunar golf in unworldly effects of one sixth of Earth gravity, and with no experience with the precise harsh conditions and topography—we believe Shepard compensated with a third ball.

Shepard was an extraordinary man who knew his great capacity—and also knew his limits. He balanced his confidence with an understanding of the situation at hand.

We can still picture Terry Pappas, in his golf clothes, heading to his own golf game, calling across the Beer Garden parking lot to Barbara, "I'd have had that third ball in my pocket."

THE RULE OF THREE

When Laura Shepard Churchley took time in her Cape Canaveral interview to reflect on her father's friendship with Carlos, she explained the relationship between Carlos and her dad as deserving of a golf ball gift. As already noted, Laura's key comments to us included her view that her sentimental dad thought in threes—and "probably took three golf balls" to the moon.

Laura paralleled a concept known as the rule of three. Why do we like things in threes? Is it because there's a past, present, and future? In Christianity, there's the Holy Trinity: Father, Son, and Holy Spirit. In art, we have the compositional rule of thirds. There are three laws of motion and thermodynamics. In theater arts, there's the three-act drama. And each story has a beginning, a middle, and an end.

Perhaps our love of the number three truly simplifies our complex reality. It's a principle captured in the Latin phrase *omne trium perfectum,* meaning "everything that comes in threes is perfect."

As mentioned in Chapter Three, documentarian Jonathan Richards purchased an old, unopened package of three golf balls dating from the Apollo era—the same brand as the ball Shepard gifted to Carlos after the flight. Like golf balls nowadays, golf balls decades ago were packaged in threes.

2. HOW WOULD SHEPARD ADD A THIRD GOLF BALL TO THE MISSION?

The manifest listed only two golf balls. To see how Shepard could add a third, we need to review the explanations by Apollo 16 astronaut Charlie Duke and other NASA personnel regarding how Shepard could easily get another ball safely to and from the moon's surface.

We already know Shepard was determined to get things right and solve problems before he left the ground. He was determined to succeed in his golfing demonstration as his last action on the moon, the pinnacle of a most distinguished career. A third ball enhanced the chance of success.

Both Slayton and Shepard knew the importance of solving potential problems *before* taking flight. We have seen that, as Blackbird pilot Terry Pappas explained the planning and thought processes of Navy-trained test pilots, such an aviator would "do it right" and "protect the mission"—in this instance completing the final demonstration approved by NASA for the Apollo 14 lunar surface work. A third golf ball would better ensure a successful demonstration of golfing in the moon's gravity. When Shepard realized the difficulty of the task he set for himself, a third golf ball improved the odds, and he could easily have gotten permission from his friend, Deke Slayton—with no justification—we learned. George Abbey, an accomplished military helicopter pilot and NASA leader, explained how easy it would be for Shepard to carry such an additional small item in the Apollo days, in his "preference kit."

Duke made clear to us, providing much detail from the days of Apollo, that if Shepard wanted to bring a third golf ball, it would be done. He also made clear how it could be done and that a few ounces transferred via the PPK—the pilot's preference kit—were easily tolerable. Duke explained that—with no question on such a simple item—Slayton would approve this for Shepard, just as Duke had received approval for the simple items he carried on his lunar mission. The manifest declaration of two balls would be no barrier, particularly in the Apollo days, with Slayton's approval. Attitudes changed after the Apollo program; in the next decade, a third ball might not have been so easily added.

These were the glory days of Apollo. World famous, Shepard remained the consummate professional aviator, attempting to solve problems before he left the ground. He would certainly have weighed the risk of a lunar golfing failure against the glory of a solidly hit ball as the world watched. He knew it would be the crowning achievement of the Apollo 14 mission.

The golf balls, as Shepard stated in more than one interview, were purchased by him, not by the government. NASA would have no objections; they wanted success as much as he did. NASA had no reason to refuse a request for a third ball. The people who approved the golf ball experiment would have every reason to want the mission to end successfully, with superb public relations appeal.

Shepard was correct in his instincts: the golfing demonstration would be one of the most memorable sports actions in world history. His brilliant idea enjoyed such worldwide success that it eclipsed his other extraordinary accomplishments. He was certainly justified in doing his best to ensure the success of the final demonstration of the moon's gravity for a worldwide TV audience.

3. WHO WOULD BE A LOGICAL RECIPIENT OF SUCH A REMARKABLE GIFT?

We must address a key factor in flown lunar memorabilia and gifts: the relationship of the astronaut and the recipient. Robert Pearlman emphasizes the key role of relationships in gifting of flown memorabilia, especially with lunar memorabilia.

Duke said he would not be surprised if Carlos and Alan Shepard had such a close friendship that "Alan would have given him some memento of his flight and the golf ball would have been the number one choice, I think."

An Annapolis graduate and Air Force pilot, Duke understood the bonds between military veterans, including military barbers and their military astronaut customers. Carlos must have enjoyed a special relationship with Shepard to justify the gift of a flown-but-unused third golf ball.

Did that close relationship truly exist?

George Abbey also was aware of the great friendship between Carlos and Shepard. In his interview, he said: "Carlos was more than just a barber. He was a very close friend."

Would this warrant a gift like an extra golf ball from the moon? Abbey was clear in his opinion: "Carlos and Shepard were very close. They did have the relationship," adding that "Carlos had Al Shepard's golf bag." (Carlos confirmed to us that Shepard gifted the golf bag to him when the two friends were cleaning Shepard's garage.)

When Carlos and Shepard played golf, Carlos explained, they would chat, but Shepard did not want to interact with the other two players if there was a foursome.

Carlos noted: "He and I could laugh, have a good time," but Shepard was "standoffish" with the other players.

Duke also spoke to special bonds between astronauts and gift recipients, including the bonds of military veterans and astronauts and their barbers. From his home in the Texas Hill Country, he visited with us about the days of Apollo and shared his personal experiences post-NASA, including the strength of the connection among veterans:

> If [Carlos] had been in the Navy, that would bring a special bond. Of two guys [former military] barbers I recall ... I think it was three or four months before one found out what my background was. You know, I don't walk around with a big sign on [saying] "I'm a hero astronaut."
>
> So I go to the post and the clerks at the post office in New Braunfels know my background. And we become good friends [thanks to the military connection].
>
> I've got a couple of guys like that at church. They're former military—probably my best friend is younger [and] was Navy, worked for the Blue Angels. And we love to hunt together, and we just have the same military background, so we meld in really well, and I can see how that could happen with Al and Carlos both being in the Navy.
>
> [Then there are] two or three other guys at our church. One is a former Green Beret and we hunt together. But really the close ones I've gotten to know from church, we just seem to hit it off. We go to Bible studies together.

Laura Shepard Churchley, too, understood the Carlos-Shepard relationship was special enough to warrant the gift of a lunar golf ball—and as we noted earlier, she believes her dad probably took the Carlos ball to the moon: as she said, "Daddy and Carlos were that good of friends."

Laura knew about Carlos the barber, and the happiness she observed in her dad when he'd been to see him about every three weeks. We believe Laura; she knew the lifelong, meaningful relationships of trust and respect NASA folk established and sustained.

Carlos's deeds were well known in his community, with his

service in public office, fighting corruption, and his great charitable actions.

The more famous man, Shepard, was lesser known for the quiet, good deeds his daughter revealed. Was he capable of a kind gesture to a friend in making use of a spare ball he might never wish to confess he had carried as an extra backup? We believe so. The special relationship between Carlos and Shepard explains the astronaut gifting the autographed ball—quietly and with no fanfare, and why Carlos never questioned the great astronaut about whether the ball flew.

This isn't just any barber-client relationship. These two men shared much—love of country and service in defense of it, preserving, protecting and defending their fellow Americans, and bettering the conditions of their fellow man. They were good friends who trusted, respected, and helped each other. They listened to and learned much from each other.

One of the world's most famous men helped—and received help—from a man who seemed on the surface to be of modest means.

There are brilliant, helpful, kind people who aren't famous at all. They surround us. We just need to learn to recognize them for their greatness. Knowing Commander Shepard and Carlos a little better, you may see why those who didn't know the celebrated Rear Admiral Alan Shepard mislabeled the complex man as "icy."

Those who knew Shepard best recognized he was capable of sentiment where appropriate and of friendship combined with perception. Shepard also recognized the discretion and trustworthiness of Carlos, who never breached confidences. On a practical level, Shepard also recognized the business relationships Carlos could make, grow, and keep. The astronaut acknowledged Carlos's experience, expertise, and abilities in the beer business.

Importantly, Shepard, it seems, was an excellent judge of character and recognized the greatness of Carlos Villagomez as a humanitarian and lifelong friend, just as the modern NASA community knows and respects Carlos's longtime kindnesses and good deeds for his greater community.

We now know that Shepard and Carlos built a relationship that developed and deepened based on their competence, skill, diligence, and trust. Both men knew the importance of discipline from their unusual early family lives.

They worked together, particularly in the year before the Apollo 14 flight, to modernize and improve Shepard's appearance into what both men preferred calling a banker's or businessman's haircut. Shepard's hairstyle—covered by the national and international press—was carefully brainstormed with several experiments in parting hair and cuts, with the best option then selected and implemented with an agreed-upon plan for training and maintaining good results over years to follow.

This is no tale of an unlikely friendship. This is very much an American story of the intersecting lives of service of Alan Shepard and Carlos Villagomez, two men from extraordinary families and American upbringing, each with a strong work ethic and dedication to learning and education, each supporting and bettering himself, his family, friends, and community and the greater world, while leaving a legacy of distinguished, courageous service to his country.

They chose the same branch of service—the Navy.

They enjoyed the same sport—golfing.

They certainly shared an interest in the business of beer and drinking. Each succeeded in business with the same entrepreneurial spirit we Americans value, with a particular love of Hispanic culture, including the language.

Each chose quests, took calculated risks, and took pride in his work, and succeeded beyond measure.

It's no surprise, then, that the two men with so much in common were lifelong friends.

We know the night before Apollo 14 isolation, Carlos came—at Shepard's request—to the astronaut's home to cut his hair. The local *News Citizen* newspaper reported the event during the Apollo 14 mission.

Robert Pearlman felt such an extraordinary gift—a golf ball from a legendary moment in sporting history—would require a special relationship, such as family member. The family, friends, and astronaut we interviewed believe a special relationship between the astronaut and the barber merited such an extraordinary gift. Pearlman seemed to understand Carlos's level of conviction that Shepard indeed gave him the ball; but the memorabilia expert did raise the possibility of a "false memory" concerning the close relationship.

A HAIR APPOINTMENT FROM SPACE?

Was the haircut appointment and golf ball gift as a token of special friendship all part of a false memory? Pearlman has found reported cases of individuals who had strongly expressed memories of personal involvement with famous space events and personalities, memories that were clearly disproven.

Is this such a case? We think not. This was no casual or early-stage friendship, such that a gift would be an impossibility; this was a close friendship verified by Shepard's daughter Laura. The Shepard-Carlos bond was well known in the NASA community, including at the level of Johnson Space Center directorship, George Abbey.

In our combat-hardened, bullet-dodging, nuclear-bomb-surviving Navy veteran friend, we have seen no signs of delusion, although we recognize Carlos clearly knows how to tell great stories.

Was the post-mission appointment and haircut a figment of Carlos's imagination? We believe both Carlos and his friend George Abbey, then a key NASA official, who are clear that Shepard had a post-Apollo 14 hair appointment with his friend and barber.

"I know that Al wanted to get an appointment," he told us.

"So that was clear?" I asked.

"Yeah. That was clear," Abbey replied. He told us, "Well, I heard it." And he also said, "I knew Al wanted a haircut. When Al came out of quarantine, he wanted Carlos to cut his hair." Abbey stated the reason as well: "Because he hadn't had a haircut all the time he was on the mission."

Abbey noted that "Carlos cut Al's hair all the time." Abbey said he could "confirm there was indeed a scheduled appointment. There was, and I'm sure Al confirmed it in quarantine."

Pearlman noted correctly that there's no mention of the haircut in the publicly available transcripts of the Apollo 14 mission; the medical channel is apparently not available to the public. Abbey confirmed it existed: "Yes, yes, there was such a channel. It is quite possible it happened on the medical channel."

We could not find a transcription for the medical channel; we do not know if such information will ever be disclosed. If the medical channel cannot be accessed to prove or disprove a reminder call from space, we know—from George Abbey—that Shepard made it clear in quarantine he wanted the appointment confirmed with Carlos. Separate from whether there was notice to Carlos relayed by NASA, Abbey said he knew Shepard confirmed the appointment in quarantine. Abbey said he had heard of the haircut request, and he knew of the friendship, and that the two were friends and "drinking buddies."

Separate from any confirming haircut call during space travel or during quarantine, Carlos is very clear: he got a call from NASA: Shepard was being released from quarantine and was on his way to the Holiday Inn. Carlos was already at his shop, waiting.

We believe Carlos's consistent story: Shepard came to his barbershop after release from quarantine for the long-scheduled haircut, and Shepard gave him the golf ball.

ON THE BARBERSHOP WALLS

Carlos's remarkable souvenirs demonstrate a history of the barber's genuine friendships with other astronauts. Beyond the colorful signed tributes on the walls (including a poster signed more than once by Shepard, with personal appreciative notes to Carlos), we also have the myriad of news articles quoting Carlos in the world's press for a moment in the 1970s when more than one newspaper included news that Shepard would be back in Carlos's chair after splashdown and the end of quarantine.

Carlos remains certain Shepard gave him the ball as soon as the astronaut appeared for a haircut in anticipation of a major international tour, meeting dignitaries and press, attending parades and parties, making speeches, and chatting to notables.

The *News Citizen* and *The Dallas Morning News* both interviewed Carlos and published articles about the future Shepard haircut appointment. Clearly, during the Apollo 14 mission, Carlos spoke to the press, who reported both the haircut at the astronaut's home before the mission and that Shepard would come back to Carlos for a haircut when he returned to earth.

Both the local News Citizen "Astro-follicles" article and one of two Dallas Morning News articles delivered basically the same story.* The Dallas paper reported Shepard's upcoming haircut appointment with Carlos, scheduled for "less than three weeks from now when quarantine is lifted."

And the world press picked up the story of Carlos the barber and his famous client's hairstyle after United Press International published it.

The press accurately portrayed the efforts by Carlos and the former Navy fighter pilot to grow out his military buzz cut successfully without looking unkempt. As we have seen, Carlos had

---

* Villagomez, press clips, February 10, 1971

worked up a longer, specially developed cut, styled with Dippity-Do, baked under a dryer, and skillfully brushed out.

How do we explain the absence of photographic proof of Carlos's hairstyling for Alan Shepard?

We credit the absence of such photos to caution by both Carlos and Shepard to maintain privacy in the details of the hair-net/bubble dryer process. Clearly, folks knew Carlos was Alan Shepard's barber, even before it made world news.

Carlos cared deeply about the quality of his haircutting and styling and about the privacy of his friend and customer. He spoke with a professional's pride when he told us of the haircut at Shepard's home before the pre-flight isolation and Apollo 14 mission to the moon that he made "sure it had a line here and a line around the ear," and that his client left the details to the pro.

Shepard knew he would be scrutinized when he began the post-flight tour NASA had planned. Carlos cared—more than Shepard at that point in time—that Shepard's shorter cut for space should also be as good as possible, even if few people would see it.

In talking about barbering, Carlos is a consummate professional, a perfectionist in haircutting. Carlos still cares deeply that this—and every haircut—be done properly. Shepard, however, left the details to his barber friend. As Carlos reminisced: "When I went over to his house, he says, 'Carlos, cut it off.'"

The result, according to Carlos, that evening before quarantine to fly in space: "he liked it."

We do not know the specifics or timing of the haircut versus the photographs taken after the astronauts were released from quarantine. At the time of Apollo 14, Carlos still barbered out of the Holiday Inn room. The press, too, was based at the Holiday Inn, but they covered the quarantine area at NASA to photograph the astronauts. Pearlman notes that the press pursued astronaut Alan Shepard.

We have no details about how NASA or the press managed the quarantine release, other than photographs showing Shepard looking handsomely groomed and dressed. We certainly know Carlos received publicity—as barber to Shepard—during the Apollo 14 newspaper coverage. While Carlos thought a phone call from space had triggered interest and publicity, it appears the press picked up the story thanks to interviews directly with Carlos. During the Apollo 14 mission, Carlos spoke openly to the press about the post-mission haircut already scheduled with Carlos by Shepard.

A few U.S. outlets personalized the story with interviews and local reporting—from inside the barbershop—while the United Press International story spread worldwide, as Pearlman noted from the avalanche of articles one can find. The articles pictured a handsome Shepard, careful about his hair, via a charming and relatable human interest story of a popular hairstyle—and hair stylist.

ASTRONAUTS AS MOVIE STARS

In the madness of early press coverage of Alan Shepard, the media covering the astronaut and his family eased up when his wife posted a note on the door of their home. In 1961, the press patiently awaited her promised statement after his successful Freedom 7 Mercury mission.

So how did the Shepards elude the press in 1971—after Apollo 14—to arrive at the Holiday Inn parking lot without being seen? We believe NASA press man Gene Horton protected the Shepards —as diligently as he had done back during the *Freedom 7* flight in 1961.

We believe Gene Horton and Alan and Louise Shepard were resourceful folks who could have engineered a solution to solve the astronaut's need for a private haircut with Carlos, just as we believe Shepard could wear a comfortable old flight suit in quarantine, and step away for a haircut across the street from NASA.

Horton was good friends with Carlos, as well. We think it's possible he suggested the hairstyling angle as a human-interest story to the press, putting Carlos on the map. We also know Horton did not discuss this Apollo 14 episode in his memoir *Losing Them*.

Horton clearly states that he effectively protected Alan and Louise Shepard from the press in 1961, after Shepard emerged as the first American in space, and that the couple would soon visit the White House and beyond. In 1961, Horton strongly believed that Shepard and his wife deserved privacy as they were being reunited and would soon meet with President and Mrs. Kennedy and headed across the Atlantic.

Now, in 1971, the couple now would meet Carlos right outside his door. Shepard could either wear street clothes under the flight suit, if time was short before meeting the press, or he could find a place to freshen up after the haircut, ridding him of seven weeks of messy hair. It makes little sense to freshen up before a significant haircut and styling; it makes sense to find a place to clean up afterwards.

Shepard insisted on privacy for his haircuts and styling. And Carlos is clear that Shepard was bent on avoiding photography during his hair styling, and hated the needed use of a ladies' dryer, a hairnet, or even a mention of the words "hair spray."

Shepard took great care to make only well-groomed appearances in public. The photographs of him that day, which Robert Pearlman indicates were taken after release from quarantine at NASA, show Shepard handsomely dressed and groomed. We do not know just how he avoided the press or how he and his wife got to the barbershop without being photographed. We also do not know the timing of Shepard's return to NASA.

Perhaps the press, clearly aware NASA could help or hurt them in future access to the astronauts, did not seek photos that would portray the astronauts in unattractive positions. Perhaps NASA allowed each of the astronauts time to see their families, attend to their appearance, and prepare before they met the press.

The bottom line: we don't know how Shepard and his wife made it to Carlos's little Holiday Inn shop and back to the Manned

Spacecraft Center across the street. We do have one history-based idea. Perhaps, in 1971, Horton—as he helped the Shepards to prepare for their visit to the White House—helped the couple again to maintain some dignity, as Alan and Louise Shepard, continued his Apollo 14 mission duties to another national and international tour.

We do not know the specifics and can't ask Gene, who died in 2021. But Shepard was recognizable. We believe he chose a comfortable flight suit as he drove with his wife to a Holiday Inn parking lot that would allow him direct access—he could simply step directly to Carlos's hotel room-turned-barbershop from the parking lot. There was no need to parade through the lobby.

We know Shepard planned carefully; he and his wife knew he'd need to be at his best after quarantine was lifted. The Shepards were more prepared than many other couples for massive attention and fame; they'd undergone the life-changing publicity and experience with John and Jacqueline Kennedy in 1961.

During the Apollo 14 coverage, national newspapers had reported that Carlos and Shepard had developed and implemented the then famous Shepard Shag hairstyle, about a year prior to Apollo 14. As outlined in Chapter One, the San Francisco Chronicle had reported that Shepard would fly from Cape Canaveral to Houston for Carlos's six-dollar haircuts.[*]

The Dallas Morning News reported—on February 10, 1971 (a day after splashdown and safe arrival home to Earth, and a month after pre-launch isolation had begun on January 11)—Shepard was scheduled for a post-Apollo 14 haircut in less than three weeks, when quarantine was finally lifted.

"Alan Shepard will be sitting in Villagomez's chair, less than three weeks from now, when lunar quarantine is lifted"[†] The Texas paper also noted Carlos was in his "small shop across from the

---

[*] Villagomez, press clip, San Francisco Chronicle, February 5, 1971, p. 9
[†] Villagomez, press clip, The Dallas Morning News, February 10, 1971.

Manned Spacecraft Center in Houston" and Shepard's barber had watched the splashdown on TV.

The *Los Angeles Times* and *Washington Post* covered similar news of Shepard's barbering of the stylish astronaut.* Shepard clearly understood that his every feature would be scrutinized during the remainder of his Apollo 14 mission.

We have no photos to prove the Carlos-Shepard haircut around the time of Shepard's release from Apollo 14 quarantine, but we believe Carlos's vivid memory of the event and Abbey's understanding: Shepard wanted Carlos to cut his hair, and the haircut happened.

THE BARBER AS MISSION SUPPORT

Shepard knew he had a significant continuing mission after splashdown, including debriefing. Duty would also require he appear on the world stage after nearly seven weeks of Apollo 14 pre-launch isolation, the mission proper, and then post-splashdown quarantine.

Apollo 14 mission isolation for the space travel began on January 11, 1971, and ran through liftoff on January 31 to splashdown on February 9, to leaving quarantine on February 27, 1971. After quarantine and their February 27 release, we know a remarkable multi-week tour lay ahead for the Apollo 14 astronauts and their wives.

They would travel to the White House to meet the president and dignitaries; to Congress for speeches before a Joint Session; to Chicago and New York for motorcades and receptions; to Florida and the Kennedy Space Center to thank the employees who made it possible; and across the Atlantic to the Paris Air Show, where they'd meet with Soviet cosmonauts in attendance.

---

* Villagomez, press clips, *Los Angeles Times*, February 5, 1971, p. 1R; *Washington Post*, February 6, 1971, p. E4.

Carlos, the enlisted Navy man, served and completed a very specific mission: to help Shepard prepare and to stand by for Shepard's release from Apollo 14 quarantine and immediately get him ready to appear on the world stage.

We believe Shepard, the superb planner, ensured Carlos's availability; he allowed his barber friend to share the excitement of the mission. In bringing his barber much closer, Shepard could also use his friend to practice for press and public questions. Carlos served his Apollo 14 mission roles perfectly, leading to an unforgettable evening in the Shepards' living room, and under the stars, featuring both a haircut and practicing for the press. Carlos will never forget the night before Apollo 14 isolation, as Shepard guided him through the nighttime sky and the landing spot on the moon.

Shepard relied on the younger barber for his discretion and for his practice questions. The astronaut could relax, with no worry about maintaining any public appearance during the haircut. Shepard was not focused on the haircut details underway at his home; he relied on Carlos to provide him a "short" haircut and then, many weeks later, to be instantly available to restore the astronaut to the quality of appearance Shepard required.

As expected, Shepard carried out his public duties handsomely; he and Carlos ensured the astronaut looked his best to perform his public role as Shepard finished his Apollo 14 mission obligations.

We assume barbers were available for astronauts in isolation, as Robert Pearlman explained. We think Shepard, meticulous about his public persona and appearance, would not trust his difficult hair to a barber other than his own, even if the alternate barber was highly qualified.

NASA certainly never contested the wave of popular stories about the Shepard Shag, or the barber whose client's job put him on the map worldwide, allowing Carlos to grow his business. Shepard benefitted Carlos; and Carlos was a loyal friend as well, sharing his insight on local businesses and personalities, traveling rural Texas as bouncer and friend, and demonstrating how to make friends in Texas bars before trying to sell Colorado beer.

A MOON DUST TEST?

It is our intention to honor the unspoken but clear agreement between Carlos and Shepard that they would not resolve the provenance of the ball after the time it was gifted—and the fact that Carlos never asked, and Shepard offered no indication of whether it flew to the moon and back.

Independently, we feel testing for moon dust would give us no irrefutable answer. It could even produce a false positive or negative, particularly since Shepard brought it to Carlos after a long stay in the quarantine of dusty moon travelers and their effects, and it wasn't handled with gloves until Carlos gifted it to Ed. We have not investigated why the ink turned pink. We see no need to violate the implied Carlos-Shepard pact that never resolved formal provenance of the ball—which could only be answered by asking Alan Shepard himself.

To our knowledge, no one ever questioned the astronaut on the topic.

At the core of the golf ball story, we must address the cold, hard fact that historian Robert Pearlman is correct: there's no signed provenance. Importantly, such proof is needed to protect everyone involved in commercial trade in memorabilia. We understand written provenance provides reliability and diminishes the risks of cheating a customer and risking the honor of the folks involved. We want to avoid such risks.

Carlos, in his interview for the documentary, says he always assumed the ball flew. He never asked Shepard, ever, for proof: "He just gave it to me, you know, and I just knew I had something good. I figured it out. But it wasn't until everything settled down that I looked into it and ... I never asked him, 'Did this golf ball go down with you to the surface of the moon or did it just stay on Earth—did it just stay in isolation?'"

Do we believe Carlos's story that Alan Shepard gave him the golf ball at the hair appointment? Yes, we love and trust the man. We believe Shepard gave Carlos the golf ball at a hair appointment.

But are we objective? No, we cannot be objective on the story of the golf ball or its history, given our love for this man and his generosity and kindnesses to us. He was and is so much more than a family barber to our late parents, to us, our children, and our grandchildren. How could we not believe Carlos, who dropped everything one beautiful sunny day to bring his razor, shears, and cape to Barbara's parents' home to barber and shave an ailing Matt Radnofsky in the backyard?

How can we not believe the word of the man chosen to carry the news to Ed of the untimely deaths of his parents? We regard him as a beloved elder, a friend to celebrate and worry about and help however we can.

He has given Ed the golf ball, and we feel an obligation to ensure his story is told, honoring the memory of the friendship between Carlos and Alan Shepard and the memory of the good works of both admirable heroes. Rather than sell the ball gifted by Carlos to Ed, we'd like to see it in a museum, used for education about the people who supported NASA and enabled the high-profile astronauts to do their jobs, just as Duke champions NASA workers—and praises old military barbers—in his interview.

The relationship between Shepard and his barber represents all the beautiful, human aspects of NASA and related personnel—thousands upon thousands of people who supported the manned space program.

How can we not embrace the opinions and explanations of Duke, who began and ended his interview by taking the time to reflect on the memory of Matt Radnofsky, with stories of Matt's and his Crew Systems' folks and their much-appreciated efforts for and devotion to the astronaut corps and to flight safety?

The youthful-appearing Duke, surrounded by books and the occasional reminder of his space travels, spoke to us at his home about the carrying of small objects in space travel, particularly to the moon. He began by asking for a moment to tell Barbara how much he valued Matt and Crew Systems. The kind astronaut, tenth man on the moon, took time to reminisce about the importance of NASA personnel, who would never be front page news.

How can we children of NASA not rely on George Abbey, renowned for his prodigious memory, holding key jobs leading him to service as the Johnson Space Center director, who worked for NASA for thirty-seven years, overseeing astronaut selection from 1978 to 1987? After our interview in 2023, with recording still underway, he spoke of the work of Crew Systems, and the extraordinary personnel, including Matt: "He played a big role. Particularly after the Apollo fire, when we were developing materials ... we used to fabricate everything .... He really contributed to our success. We need more people like him today."

SHEPARD STANDS OUT; CARLOS DOES, TOO

Duke explained Shepard's unique and immense role in the history of space flight and singular dominance in the history of space exploration. Shepard stands out.

It won't surprise you, then, that after our interview with Duke was over, he asked about seeing the Carlos golf ball. He picked it up from the case Ed used to carry it.

Duke examined the ball, the autograph, inscription, and brand, and then exclaimed: "Well, now we know the brand of the other two on the moon's surface!"

Loving Carlos, how could we not agree with Laura Shepard Churchley as she spoke of her belief her dad took three balls to the moon and brought back one he gifted to Carlos?

We agree with Laura that the two men were that close of friends to warrant Shepard bringing home an unused golf ball from the moon and quietly, with no fanfare, signing and gifting it to his friend, the barber.

We remain your biased authors. Carlos remains our dearest friend from our early years. We agree with his long-held assumption: the Carlos ball flew.

We find it a fine tribute to friendship that Carlos never asked for proof, and Shepard never offered written evidence the golf ball went to the moon and back.

We understand the importance of written provenance to collectors and people like Robert Pearlman, who serve in the business of space memorabilia, evaluating provenances, as valuable souvenirs are bought and sold. A written provenance protects everyone involved in commercial trade of the items at issue.

Clear, written historical proof helps avoid fraud, which could cheapen or destroy the value of memorabilia. Proper provenance provides reliability and diminishes the risks of cheating a customer, destroying the honor of those involved in the process. Proper provenance is a genuine concern to anyone wondering what the golf ball's value is, what it might fetch at a sale. We regard the Carlos ball as an emblem of friendship between an astronaut and his barber, both serving NASA and their greater community, in historic times. That's priceless.

We believe Pearlman's argument of "no written proof" also speaks dramatically to the relationship between the two men.

It sufficed for both Carlos and Alan Shepard that Shepard gave him the autographed ball, at a time and in circumstances that made the gift most special. Had Carlos requested and received a signed testament of flight, we would never have understood the special friendship, or the special qualities Alan Shepard and Carlos Villagomez held in common.

To us, it's enough that Carlos and Shepard decided by their silence over the decades that there would be no request for proof and no offer of written provenance. Neither man spoke about written proof being given from one to the other. Carlos knew his friend had grinned and agreed when Carlos asked him to "do something for me up there."

We believe Shepard did indeed do something for Carlos and brought the golf ball back from "up there."

WHAT'S IN A NAME?

We need no crystal ball to predict the two balls left on the moon by Shepard are MaxFli—the same as the brand on the Carlos ball. We believe Shepard's choice of brand name reflected his great sense of humor, described by both his daughter, Laura, and Carlos. Shepard made clear in an interview from the 1970s he purchased the golf balls himself.

Laura Shepard spoke to us lovingly about her dad's "fun sense of humor." Carlos made a similar statement during Apollo 14 to the press about Shepard's fun side, before Carlos even knew about the golf ball or the planned-and-approved golfing part of the Apollo 14 mission. Recall Carlos's comment to the press about his friend: "He's a cat, man. He drives real fast cars. He dresses sharp. He's an up-to-date fellow. He's got a great sense of humor."

We believe Carlos's calm conviction the golf ball flew will be proven accurate someday. In his 1981 interview, Shepard told the world he hoped for future returns to the moon for further exploration and mused that future lunar explorers would play with the two golf balls he left behind. (Shepard, Academy of Achievement interview, February 1, 1981).

The Carlos ball—MaxFli, sold in traditional packages of three—would have two siblings, so to speak. As explorers return to the moon and locate the two golf balls there, we expect their brand name to be the same as that of the Carlos ball, the cleverly appropriate name "MaxFli."

We give great weight to Duke's explanations of how straightforward it would have been to for Shepard—or any of the Apollo astronauts—to ask permission from Deke Slayton to take a small additional item, and the ease with which a small item such as one golf ball could have been added as part of the pilot's preference kit to be taken to the moon's surface.

As Pearlman reasoned, if Shepard had hit the first ball well, that would have been the end of things. But he missed the first, hitting the second. Those two balls are still up there, Shepard explained. What if Shepard had also whiffed the second ball? We believe Shepard, a stickler for preparation, planned for that possibility by bringing the third ball from a traditional "Max-Fli" package he purchased. Happily, a third golf ball proved unnecessary for the demonstration.

We hope the Carlos golf ball will be viewed as a wonderful memento, a reminder of the people behind the astronauts, who also played key roles, as appreciated by astronauts and the NASA personnel we were privileged to meet and interview, thanks to the golf ball gift.

And if the third golf ball is not the Carlos ball gifted from Shepard, then where is that third golf ball?

CHAPTER 6
FOR THE FUTURE

THE JOURNEY of exploring the golf ball gifted by the astronaut to the barber brought us to important territory. We learned about the quiet and extraordinarily important philanthropy of America's first man in space. Alan Shepard began the Astronaut Scholarship Fund with the surviving Mercury astronauts and Gus Grissom's widow, Betty Grissom. It is now called the Astronaut Scholarship Foundation, with leadership including Laura Shepard Churchley and Charlie Duke, together with many other generous astronauts and supporters.

We first learned of Shepard's legacy from Duke at the close of our interview. The Astronaut Scholarship Foundation uses astronaut artifacts to fund key science scholarships. Treasured artifacts from early space flights, however, faced threats in the past. Ed is now a mentor to a young scholar for the organization.

Some of the historic figures we interviewed—unbeknownst to us as we were setting up interviews—were involved in efforts to preserve rights to astronaut mementos. Their organized sales of such items benefit science, technology, engineering, and mathematics (STEM) scholarships for deserving students.

The Astronaut Scholarship Foundation seems laser-focused,

fulfilling the ongoing educational mission of Alan Shepard. The foundation has now grown significantly, thanks to its leadership and generous donors, including astronauts.

In our view, the Carlos golf ball represents the friendship between two men who on the surface seemed to share little. But that's not the whole picture. Carlos Villagomez and Alan Shepard trusted and helped each other. They treated each other with great respect and listened and learned from each other.

One of the world's most famous men helped and received help from a man who seemed, on the surface, to be of modest means. We regard this as a lesson: there are many brilliant, helpful and kind people who aren't famous at all. They surround us. And we should celebrate them and recognize them for their greatness.

So we celebrate Shepard's gift to Carlos as a reflection of the humanity and the accomplishments of the U.S. space program. The Apollo 14 mission was the mission of a lifetime for Alan Shepard, who proposed, fought for permission, planned with experts, and ultimately gained permission for a celebration of the space program with the sport he loved.

Shepard's lesson involving golf would depend on a successful mission, concluded on time. From his joint memoir with Deke Slayton, it's clear that Shepard knew Apollo 14 was the pinnacle of his career. He hoped—and he carefully planned—for a celebration of this moment through the sport he loved. After his success, we believe he provided a memento of his mission with his friend awaiting his return.

What's next for the Carlos golf ball?

A HOME FOR THE GOLF BALL

We believe the value of the Carlos golf ball is its representation of the friendship between one famous man and one of the countless thousands of workers who streamed to a space city built on a cow pasture as part of a successful, uniquely American quest to put a man on the moon within a decade of an American president setting forth the mission.

We hope to see the golf ball displayed at a museum or other educational institution, to honor and teach about the people whose service, friendship, and love created and sustained—and still enable—the work of the Johnson Space Center and its people and their descendants. NASA pioneers' heirs carry on their work, including the people who helped us produce this book and the companion documentary.

You can still find Carlos at his Beer Garden, now run by one of his sons, or swapping stories at the Barbershop, holding court in the tiny, tidy shotgun-shack decorated with ever-changing space mementos.

He might be flirting a few steps away with the wait staff or telling stories with visiting dignitaries, NASA workers, bikers, and oil field workers having lunch or a beer. Everyone knows Carlos, it seems.

Today, he also has new and younger friends. The current crop of scientists and astronauts at NASA honor Carlos for his devotion—his calling—as a friend to the entire NASA community. A new generation has discovered his stories, the tiny barbershop, and Carlos Beer Garden.

Particularly kind friends, including Jon Powell, former mayor of Taylor Lake Village and a current city councilman, active in the Bay Area Houston Economic Partnership, helped in so many ways with the documentary film and this book. Among his contributions were meeting up with Carlos and his buddies at the local McDonald's for breakfast and visits, checking in on his status, fixing his walker, showing him how to update his phone, getting him to appointments, lunches, and on trips to see his childhood home by the Ship Channel—and setting up bird feeders at his home.

Newer friends like Jon enrich Carlos's life as he and his friends and family have enriched ours; we met Jon through Barbara's and Jon's mothers, who were neighbors in Timber Cove into their eighties. They shared common experiences as teenagers married in June 1945 to their sweethearts returning from Air Force combat service in World War II, benefitted by the G.I. Bill and devoted to science.

Jon facilitated many of our meetings, Zooms, and interviews for the documentary, ensuring Carlos's participation (including his first use of Zoom) and benefitting this book. Jon's wife, Cindy Evans, is a planetary geologist who trains astronauts in geology for their missions. She also participates in other research activities, including missions to Antarctica to explore, collect, and study meteorites—space objects that land on Earth.

Upon the publication of Barbara's first little book of Carlos stories, we celebrated Carlos with a Carlos storytelling night at the Beer Garden. One of his granddaughters is now a favorite waitress, with accomplished people and management skills. And as mentioned, one of Carlos's sons—Adolph—now runs the Beer Garden. Carlos proudly boasts also of another granddaughter who is a master chief in the Navy—the most senior position for an enlistee—based in the Pentagon and temporarily stationed at the Space Center, and who outranks him. (Carlos was a bosun's mate, third class.)

The little subdivision we knew as Timber Cove—across the lake from Ed's family home in El Lago—is now affectionately called the "Astrohood" by the locals, and is part of the much greater developed area now known as Taylor Lake Village.

The young scientist-astronauts and NASA workers, area contractors, and bikers and day laborers and schoolteachers and Rotarians all still mingle at Carlos Beer Garden, especially for karaoke nights, as old-timers sit a bit farther out at the picnic tables.

But no one has come around to replace the friendship of Carlos's younger days with his buddy Shepard.

If you ever wonder what life is about, consider the continuing journey of Carlos and his family and friends—at the shotgun-shack-of-a-barbershop and the big metal barn of a Beer Garden/Ice House facing the railroad tracks near NASA Road 1 and Old

Galveston Road. The whistles of the Southern Pacific trains warning pickup trucks and animals to get off the tracks still periodically drown out all conversation across the street at the Beer Garden. At night, way in the back, the moon still dominates a surprisingly dark sky.

If you visit, you can meet some of the everyday living heroes of aviation, the successors of Charles Lindbergh. Most are folks living near NASA or still working there are part of the past, present, or future of enabling the proper functioning of the increasingly diverse space program and industry.

There are also oil field workers, ranch hands, bikers in regalia, sometimes costumed karaoke singers, scientists, and business leaders and other larger-than-life, modest, and heroic figures, including those mentioned in this book.

Our larger-than-life heroes include George Abbey, Joyce Abbey, Jeff Ashby, Joe Allen, Mike Coats, Charlie Duke, Terry Pappas, and the heroines and heroes of Artemis, one of whom Duke hopes will soon replace him as the youngest person to walk on the moon.

We treasure their help and lunchtime visits, and continue to learn from the astute questioning and encouragement of many others who join us at the Beer Garden to listen to stories and brainstorm, including Ronnie Krist, Kim Krist, Rob Parrish, Christine Rowan, Sally Antrobus, and her kind family, several fine Rotary leaders, and the newest Webster police chief, who enters with no guns blazing, as did his predecessor in the old days, but as a friend of the respected owner. He dropped in for a recent picnic table lunch out in the back and swapped stories with Carlos. We were also joined by representatives from the Astronaut Scholarship Foundation when they came to town.

The NASA people are local heroes who include the astronauts, workers, scientists, and mathematicians and their family members and members of the surrounding community who support NASA's work and have helped advance our understanding of the universe, improving our lives. Their employers include private and public entities like Space Center Houston and the Lone Star Flight Museum (built on the Ellington Air Force Base Officer's Club pool, where Ed played as a child), educating and preserving our space program heritage—including the people in their souvenir shops who provide items that might inspire the next Alan Shepard.

Our heroes indeed surround us. You can still wander out back over the same land behind Carlos Beer Garden where Alan Shepard mingled, shouting jokes in Spanish with the day workers on ladders and up high in trees as they shouted questions downward—what it was like "up there" working and playing golf on the surface of the moon?

We've sat out in the black nights with Carlos and friends and looked up at the moon and stars and felt the connection.

ACKNOWLEDGMENTS

The most beautiful thing we learned about Alan Shepard, in preparing this book and assisting in the documentary, was that he truly was a sentimentalist. We read in history books of the great and varying talents and capacities of America's first man in space, but little has been written of Shepard's private kindnesses, tenderness, and quiet generosity to those in need—aspects of him that provide us a larger view of the man, beyond the portrait of out-of-this-world intelligence, quest for perfection and an impressive list of achievements.

The people who aided us were kind, and indeed tender, also showing a keen sense of duty in their educational, scientific, scholarly, and joyful approach to the information they conveyed. Jonathan Richards has retained copies of the interviews with people we quote here, as has Ed. Here's hoping you will find such heroes in your journeys; the intersecting journeys of Carlos Villagomez and Alan Shepard and the people referenced in these pages reflect commendable, honorable lives of service to country, community, family, friends, and self.

TO ROBERT PEARLMAN AND THE PEOPLE OF NASA AND SPACE CENTER HOUSTON

We are grateful to Space Center Houston (www.spacecenter.org) for permission to film the interview of the well-known and respected space memorabilia expert and historian Robert Pearlman; his expertise and the inspiration of an extraordinary day at NASA's Johnson Space Center and Mission Control improved our understanding and further motivated us to do our best to do right by this story.

We are indebted to the expertise of Robert Pearlman—recommended to us by esteemed space historian Andy Chaikin. Andy also kindly located old materials on Barbara's father, particularly the government orders, returning him home to Houston from Washington after the Apollo 1 tragedy. Please find more information about a few of Chaikin's many brilliant works in the Sources section at the end of this book.

Pearlman educated us and challenged us to explore and learn more about the golf ball and the issues surrounding space memorabilia.

Pearlman is a highly regarded space historian, journalist, and author and the founding editor of CollectSpace.com, a daily online publication chronicling space exploration and its intersection with pop culture. He actively engages with collectors, historians, and NASA folks; and he seemed quite a philosopher as well, speaking movingly of humankind's eternal fascination with space and space memorabilia.

Pearlman is respected and trusted as one of the first web developers to place online a space-related website not run by NASA, gaining traction as he initially sought to catalogue his own space memorabilia collection. After more than two decades, he provides more than artifacts; he explained in his interview for our documentary that he's been "collecting the stories and the people that are behind those artifacts and make the history what it is."

Pearlman has co-authored work published by Smithsonian Books and developed content for the National Space Society and Apollo 11 moonwalker Buzz Aldrin. He serves on the History Committee of the American Astronautical Society, the advisory committee for The Mars Generation, and the leadership board For All Mankind. We turned to this historian for a key interview, noting that he'd been honored by the American Astronautical Society with the Ordway Award for "Sustained Excellence in Spaceflight History" (https://www.space.com/author/robert-z-pearlman).

In founding CollectSpace.com, Pearlman has built an extraordinary community online, bringing diverse people together and pooling their knowledge. We learned he'd been interested in space since the age of six, wanting to be an astronaut. He had supportive parents, including an influential mother, a former teacher. He speaks movingly of the great education he received in school, in travel, and in exposure to the people and places involved in the sciences, the arts, aviation history, and space exploration.

We wanted to know his opinions on whether the ball flew. He gave generously of his time and charged nothing. We learned much from him, and he spurred us to learn more. We thank Rob Pearlman for helping us put the true value of the ball into perspective.

TO CHARLIE DUKE, MIKE COATS, TERRY PAPPAS, GEORGE ABBEY, JOE AND DAVE ALLEN, JOYCE ABBEY, KRIS CARPENTER STOEVER, AND JESSE SALINAS

We appreciate beyond words the patience and expertise of Charlie Duke, Mike Coats, Terry Pappas, George Abbey and Joyce B.K. Abbey, Joe Allen and his brother Dave Allen, and Jesse Salinas for explaining historical information we would never have known. These modest heroes were generous and patient with our questions and thirst for stories about the old days.

Joyce B.K. Abbey, SAIC, chief investigator for the Johnson Space Center Knowledge Management Office Case Study, provides NASA with key lessons for current and future programs through case studies on significant aerospace events. She also serves as director of Employee Communications and External Relations, and director for Safety & Mission Assurance Engineering; manages

Houston area STEM outreach; and creates aerospace education events for our community. Joyce represents the future, inspiring and educating our greater community. The daughter of longtime Johnson Space Center chief George Abbey, Joyce understands the greater community served by NASA and is a key link to the world of education.

You've read of George Abbey's kindnesses to us in meetings, using his legendary memory to provide details of the friendship between Carlos the barber and Alan Shepard, as well as offering meaningful reflections on the unsung, brilliant personnel of NASA's Crew Systems division. Independent of Carlos, our last recorded interview with George Abbey—on May 25, 2023—gave us the highest possible level of firsthand knowledge of (1) the likelihood Alan Shepard took a third golf ball to the moon, and (2) how it could have been done.

The extraordinary George W.S. Abbey passed away on March 24, 2024. The family wrote movingly: "He has been called the father of modern spaceflight, but we called him Dad, Grampa, and Uncle George. He was a quiet man, brilliant, humble, and very private. The world will be so much emptier without him."

Many of Abbey's accomplishments in his long career are described in Michael Cassutt's *The Astronaut Maker*, thanks to exhaustive research which also provides insight into Abbey's apparent feelings on protection of his family's privacy.

Joyce Abbey and her father provided insight into his personal side and his pride in bringing Texas longhorns back to the pastures at JSC. In addition to his professional accomplishments, he was a man of deep faith, fond of the Houston Livestock Show and Rodeo, classic cars, Celtic music (and attending the annual festival in Brittany), fine wines, and friendships. Joyce wrote: "He is survived by his five children, eight grandchildren, three great-grandchildren, nieces, and nephews, and by so many whose space careers he launched and nurtured."

Kris Stoever, a grade school buddy and old playmate of

Barbara's, co-wrote with her dad, Scott Carpenter, a compelling, sensitive, and scientific biography/autobiography (see Carpenter and Stoever, *For Spacious Skies*). She gave us a chance to reminisce and to thank her for her dad somehow hauling Barbara and the family dog out of the canal behind his house when she was a little kid, playing too close to the edge.

We first heard of the Terry Pappas's great tales of flying the Blackbird, Joe Allen's Shuttle exploits, and Mike Coats's wonderful, inspiring story about his extraordinary wife — and what she accomplished in her inspirational dealings with President George H.W Bush — thanks to the late master barber Jesse Salinas. He gave much of himself to his community. Jesse was planning to realize his dream of following in the footsteps of Carlos, who served as a city councilman and Mayor Pro Tem of Webster.

Jesse was an unsung hero, much more than a barber and beloved friend. Beyond his care for his loved ones, his faith, and his great barbering skills, he devoted much of his time to volunteer work in Galveston as part of a group of friends who formed the "Galveston Golf Cart Society" for helping the community. Group members conduct charitable volunteer projects, cleaning up the beach, delivering goods, and engaging in quiet, heroic efforts to make lives better on Earth.

TO ALAN SHEPARD AND LAURA SHEPARD CHURCHLEY

In working on this book, Barbara suddenly remembered a book gifted to her almost three decades ago, shortly after her dad died; never opened, it was put away in hopes for a time when memories wouldn't be so painful. It was a first edition of Alan Shepard and Deke Slayton's 1994 book *Moon Shot: The Inside Story of America's Race to the Moon*, with a surprise when finally opened almost thirty years later: a thoughtful and kind inscription.

To Barbara—In memory of Matt Radnofsky—we could not have done it without him.

    —Alan Shepard Freedom 7, Apollo 14

Just as the book's contents intensely portrayed a complex, disciplined man with a philosophical side, the inscription showed Shepard as tender and kind, specifically considering the grief of the daughter of a NASA worker whom he took time to honor in that inscription. The book's contents reinforced Laura Shepard Churchley's portrayal of her father as a sentimentalist in the best sense of the word.

Both Laura Shepard Churchley and Charlie Duke gave generously of their time and shared Alan Shepard with us in most meaningful ways. We continue to admire them for their patience, dedication, and leadership in the Astronaut Scholarship Foundation (https://astronautscholarship.org/scholarshipprogram.html). Ed volunteers there as a mentor, and we have both enjoyed supporting their important good works.

TO DOCUMENTARIAN JONATHAN RICHARDS

Jonathan Richards undertook a short film about Carlos's story, contained in Barbara's earlier book, *Listening Space*. The little book focused on stories from the barbershop told by Carlos and his NASA clients, demonstrating many of the concepts Barbara uses in teaching problem-solving to children and in her work as a professional mediator.

Jonathan is a remarkable film producer, director, and editor from the United Kingdom, traveling globally in his work of creating a variety of commercial and documentary films (https://www.jonathan-richards.tv/). As an aeronautical engineer, he worked in British aerospace as a military aircraft designer, with roles on Hawk, T45, Harrier, and Buccaneer aircraft types.

We met Jonathan thanks to our niece, journalist Caroline

Radnofsky, currently a supervising reporter for the NBC News team based in London. She worked with Jonathan as he was commissioned by a news network to film a documentary about Matt and his journalist-granddaughter's search for his history. Caroline produced and narrated the film. She learned—and taught us much—about Matt Radnofsky's military service in bombers during World War II and his work for the military, NASA, and industry contractors working for NASA.

For the earlier film, Jonathan and Caroline filmed in England, Canada, and throughout the U.S. They captured great storytelling by Smithsonian museum experts, veterans of the early NASA and major contractors' programs, and U.S. and Canadian astronauts from the Apollo days through the present.

One of Caroline's interviews featured an astronaut you've met in this book: Mike Coats. And where had filmmaker Richards captured Coats's storytelling in his earlier documentary?

It happened while he was receiving his regular haircut from master barber Jesse Salinas, in Carlos's barbershop.

And at the core of this book, we acknowledge our love and gratitude—

TO CARLOS

For his admirable life of service and loving kindness to the people we love and to so many others who have benefitted by knowing him.

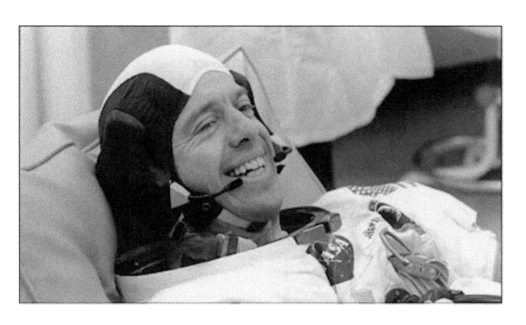

SOURCES

Abbey, George. Interview by Barbara Radnofsky, Webster, TX, May 25, 2023. Recordings 4–6 contain George Abbey. Collection of authors.

Allen, Joseph P., and Russell Martin. *Entering Space: An Astronaut's Odyssey*. New York: Stewart, Tabori & Chang, 1984. Magnificent photographs illustrate a firsthand—and first class—account by Joe Allen of what it's like to be an astronaut on a Shuttle flight.

Astor, Gerald. *The Mighty Eighth: The Air War in Europe as Told by the Men Who Fought It*. New York: Dell Publishers, 1997. Accounts of many air battles. (See also Russell Strong, *First Over Germany,* for details of the 306th Bomb Group and Lt. Radnofsky.)

Astronaut Scholarship Foundation (ASF). Website. Cited in the court case *Bean v. Bean* recognizing the rights of Mercury, Gemini, Apollo, and other early astronauts to keep their mission artifacts. ASF explains: "The Astronaut Scholarship Foundation was created in 1984 by the six surviving Mercury 7 astronauts (Scott Carpenter, Gordon Cooper, John Glenn, Walter Schirra, Alan Shepard and Deke Slayton); Betty Grissom (widow of the seventh astronaut, Virgil "Gus" Grissom); William Douglas, M.D.

(Project Mercury flight surgeon); and Henri Landwirth (Orlando businessman and friend). Together they represented a wealth of collective influence that was particularly suited to encouraging university students pursuing scientific excellence. Their mission was to ensure that the United States would be the global leader in technology for decades to come. Skylab and Space Shuttle programs, ASF partners, industry leaders, universities, and individuals have joined in providing scholarships to top university students pursuing degrees in science, technology, engineering, and mathematics (STEM). ASF has a "lifelong relationship" with each astronaut scholar. https://astronautscholarship.org/aboutasf.html

Ball, David S. *American Astrophilately: The First Fifty Years*. Charleston, SC: A&A Publishers, 2010. This history of flown aviation postal covers in the U.S. includes a contributions from memorabilia expert Robert Pearlman on the Apollo-Soyuz Test Project. The book also references Matt Radnofsky's flammability safety work on the development of Beta cloth and fireproof paper.

Bean, Alan, with Andrew Chaikin. *Apollo: An Eyewitness Account by Astronaut/Explorer Artist/Moon Walker, Alan Bean*. Shelton, CT.: The Greenwich Workshop, Inc., 1998. Astronaut-turned-artist Bean provides an interesting account of his use of "some of the tools I had with me on the moon," during his Apollo 12 mission to the moon.

Bean wrote that he kept the tools. See also *Bean v. Bean* and discussion in Chapter 4 of the present work. In his own illustrated art book the moonwalker stated that he used these tools in preparing his famous canvasses containing moondust. He references and explains his use—on his canvasses—of "my heavy, metal geology hammer—the same one I used on the moon to drive the staff that held up our American flag, and to pound in the core tubes to obtain soil samples" (154).

He also explained (155) his use on the surface of the canvas of "a sharp-edged bit that was on the cutting edge of one of the core tubes," noting that "these tools, which once helped me explore the moon, are now putting the moon's stamp on my paintings."

The astronaut/artist added that he would mix in bits of charred portions of the heat shield with gold foil used to cover the spacecraft's side and hatch. Bean received these items from Max Ary, the museum director who'd received the Apollo 12 spacecraft for display. The museum director had found, upon opening the shipping container, the tiny portions of the heat shield and gold foil, which he gathered and sent to astronaut Bean (155).

Bean emphasized that he was able to obtain moondust by using the dust from American flags, NASA insignia, and Apollo patches, all gifted to him by NASA. When he realized these framed items were dirty with Apollo 12 moondust, the astronaut/artist explained that he cut off a portion of each of the emblems, chopped the portions into small bits, and put a few of the bits into the modeling medium. He concluded: "I have found a way of putting a little bit of the Ocean of Storms into each of my paintings" (155).

Bean v. Bean, No. 05-21-00286-CV, 2022 Tex. App. LEXIS 9058; 658 S.W.3d 401 (Tex. App.—Dallas Dec. 13, 2022, pet. denied).

Brown, Brandon R. *The Apollo Chronicles: Engineering America's First Moon Missions.* New York: Oxford University Press, 2019. Son of an Apollo engineer, Brown includes a brief history of spacesuits (92). He focuses on storied engineers, including Max Faget of Newport News, Virginia, who engineered and refined key elements, objects, and aspects of the space program. Faget moved his family to Dickinson, Texas, in the 1960s to build a home near the Manned Spacecraft Center. His daughter Ann reminisces that her father designed a long hallway with a golf hole covered with a plug and carpet, where her dad would practice his putting (96). The book also speaks to the benefits from active experiments left on the moon surface (236) and the tendency of NASA personnel to accumulate souvenirs, in the chapter "Today—Mementos and Returns."

Carpenter, M. Scott, Gordon L. Cooper, John H. Glenn, et al. *We Seven: By the Astronauts Themselves*. 1st ed. New York: Simon & Schuster, 1962. This famous and popular book, an obligation under the seven original Mercury astronauts' exclusive contracts with *Life* Magazine, provided contributions to education and history, spurring public interest. One of Shepard's essays, "What To Do Until the Ship Comes In," explained the importance of "down-to-earth hard work," and the grueling training for being prepared in any situation for emergencies and whether they ditched on land or sea. They trained for even the slim possibilities, he explained, "...we did not want to leave anything to chance." (204). The astonauts also provided fascinating essays in an earlier, lesser known 1961 book (Cooper, et al *The Astronauts: Pioneers in Space*) about their expectations for Mercury and the future, with Astronaut Cooper providing the first account of their plans and thoughts. Shepard's brilliant, short essay in the earlier book captured his approach to containing fear – via relentless training and preparation.

Carpenter, Scott, and Kris Stoever. *For Spacious Skies: The Uncommon Journey of a Mercury Astronaut*. Orlando, FL: Harcourt, 2002. 1st ed. This is a professional explanation for controversies surrounding Scott Carpenter's Aurora 7 Mercury flight (265–305) within a moving account by Carpenter and daughter Kris Stoever of the astronaut's challenges, including tending a sick mother and being derided by his straying father as a "flying bus driver." Carpenter provides intensive detail about the simulator and praises CapCom Shepard as an experienced, insightful, helpful, and kind colleague/aviator at key moments. The Carpenters lived near the Radnofskys, in Timber Cove, where Mr. Carpenter once fished Barbara and the family dog out of a canal behind his house.

After the astronaut's Aurora 7 Mission, Mr. Carpenter expressed handsome gratitude, written on a large poster-sized blowup of a NASA photograph of the astronaut working with a Matt Radnofsky invention. The NASA-framed photograph bears this handwritten description, signed by Carpenter: "For Matt, with best regards and many thanks to the designer," in gratitude for one of Matt's inventions: a special life raft that the astronaut put to very good use as he calmly awaited ocean pickup in the raft after the Aurora 7 splashdown, an episode he details in his book.

Cassutt, Michael. *The Astronaut Maker: How One Mysterious Engineer Ran Human Spaceflight for a Generation*. Chicago: Review Press, 2018. A meticulously researched portrait of George Abbey and NASA history, with a title understating the qualifications of the subject as well as the historical scope of space program personalities and heroes depicted. In his 90s, throughout multiple lunches and in our interview—outside in the heat in back of Carlos Beer Garden—George Abbey displayed his famously prodigious memory and sense of humor.

Cernan, Eugene, and Don Davis. *The Last Man on the Moon: Astronaut Eugene Cernan and America's Race in Space*. 1st ed. New York: St. Martin's, 2000. Cernan, who had a distinguished career as pilot of Gemini 9, lunar module pilot aboard Apollo 10, and the moonwalking commander of Apollo 17—wrote that, awestruck at meeting the "Original Seven," he had each of them sign his copy of Carpenter et al., *We Seven*. Once Cernan joined the program and assessed astronauts as possible flying partners, he gauged Alan Shepard "the best of the bunch" (65). He viewed Shepard's "icy-cold and abrasive" nature as a benefit and concluded that the good cop/bad cop roles played by Slayton and Shepard instilled confidence and "demanded more than your very best. The result was a better program" (67). Cernan and your author Ed flew as private pilots out of the same airport and hangar. The great old test pilot was friendly and most generous with his time.

Chaikin, Andrew. *A Man on the Moon: The Voyages of the*

Apollo Astronauts. 1st ed. New York: Penguin Books, 1994. A classic from a most respected space historian, deeply researched and annotated in detail, based on firsthand accounts, including the authors' interviews with 23 moon voyagers.

Chaikin, Andrew, and Editors of Time-Life Books. *A Man on the Moon*, vols. I, II, III. Alexandria, VA: Time-Life, 1999. The three volumes expand on the meticulously researched original Chaikin work with historical context, photos, and references to popular culture.

Churchley, Laura Shepard. Interview by Jonathan Richards, Cape Canaveral, FL, April 17, 2022. Video interview. Collections of authors and documentarian.

Coats, Michael. 6 interviews. NASA Johnson Space Center Oral History Project. The interview of November 9, 2012, tells the handsome story of his extraordinary wife changing history with the suggestion of a small, flown gift for First Lady Barbara Bush: a tiny souvenir shuttle charm. Comander Coats patiently reviewed and provided much additional information for the stories he provided us in this present work.

https://historycollection.jsc.nasa.gov/JSCHistoryPortal/history/oral_histories/CoatsML/coatsml.htm

CollectSpace.com. Website of respected memorabilia expert Robert Pearlman, a web pioneer with one of the early space-related websites not run by NASA. He provides historical information on artifacts, including documents and first person accounts of the artifacts and the folks behind them. The website, for example, contains NASA Apollo era policies and procedures for astronaut personal preference kits, when last visited in late 2023.

http://www.collectionspace.com

Collins, Michael. *Carrying the Fire: An Astronaut's Journeys*. New York: Farrar, Strauss & Giroux, 1974. Beautifully written U.S. astronaut memoir introduced by Charles Lindbergh.

———. *Flying to the Moon and Other Strange Places*. New

York: Farrar, Straus and Giroux, 1976. A small, appealing book, explaining scientific principles in understandable language and diagrams, combined with more personal perspectives on flying to the moon: "I will never forget how beautiful the earth appears from a great distance, floating silently and serenely like a blue and white marble against the pure black of space. For some reason, the tiny earth also appears very fragile, as if a giant hand could suddenly reach out and crush it" (Chap. 12).

Cooper, Major L. Gordon, et al. ("the seven astronauts of Project Mercury"), with Loudon Wainwright of *Life* Magazine. *The Astronauts: Pioneers in Space.* New York: Golden Press, 1961. An inspiring and richly illustrated work provides an essay by each astronaut as they prepared for the Mercury project and as excitement built in the country. Alan Shepard's short, science-filled essay began with a concession of fear when flying a night mission off Korea, as his radio failed in heavy overcast and knew he might need to ditch in bad weather at night. "My training saved me," Shepard wrote. "Years of learning that the best way to treat a problem is to concentrate on alternate methods of solving it made me able to hold the fear back."

Cunningham, Walt, and Mickey Herskowitz. 1st ed. *The All-American Boys.* New York: Macmillan, 1977. The Apollo 7 astronaut provides detailed stories and critique of the U.S. space program, NASA management, and astronaut colleagues. From the outset, Alan Shepard appears as "most competent" (18); respected for his return to conditioning and preparation for Apollo 14 (231); and admired by "even the most cynical" of astronauts for "the sophistication and social poise he brought to the job" (36). Shepard shines in his astronaut office responsibilities and successful business judgments and ethics (171–73), rising above commercialization issues.

Diaz, Jaime. "Shooting for the Moon," *Sports Illustrated,* August 3, 1998. The son of a country club pro claims his dad supplied the golf balls used on the moon by Alan Shepard, as *Sports Illustrated* airs the debate, including conflicting views by Shepard initially and later as to whether the golf balls are well-preserved or not in the moon's harsh conditions.

Duke, Charlie. Interviewed by Barbara Radnofsky and Jonathan Richards, Duke home, Texas Hill Country, April 16, 2022. Video interview. Collections of authors and documentarian.

Engle, Eloise, and Arnold Lott. *Man in Flight: Biomedical Achievements in Aerospace.* Annapolis, MD: Leeward Publications, 1979. As explained here, the "Apollo 14 crew was the last to be subjected to quarantine precautions. Because a member of the Apollo 13 crew had been exposed to a communicable disease, a special program was set up for Apollo 14 designed to strictly limit their contacts with outsiders before the flight; only wives and about 150 people considered essential had any direct contact with either prime or backup crews" (352).

Gilbreath, K. B. Oral History. From White Sands, Utah to Webster, Texas, Mr. Gilbreath undertook the massive tasks of handling the development and maintenance of NASA's facilities and roads, requiring much thought and flexibility in accommodating new needs and new technologies.
https://historycollection.jsc.nasa.gov/JSCHistoryPortal/history/oral_histories/GilbreathKB/GilbreathKB_2-26-03.pdf

Green, Bill. Email correspondence to Barbara Radnofsky, March 22, 2024, regarding meeting Alan Shepard in 1964. "As a fan of the space program I got the chance to visit Houston with my family in December of 1964, during the Gemini program. We stayed with my parents' dear friends the Radnofskys, who lived near the Manned Spacecraft Center. Matt Radnofsky was chief of Crew Systems for NASA. He worked with the astronauts in our attempts to beat the Russians to the moon. He asked my dad if he would like to see where he worked and [my dad] said he would. He asked me the same question and before I responded I was out the door and already in the car!

"We drove from Timber Cove to the MSC, which probably took ten minutes but for me it seemed like a year. We pulled into the parking lot, exited the car and started our walk. It was Matt, my dad, my sister Joanne and me. As we approached the doors of the building, here comes Alan Shepard, the first American astronaut to go into suborbital flight. The first thing that I noticed was his grin, which seemed to be as wide as the distance between the earth and the moon. At that point I was a 14-year-old thinking that this wasn't really happening, and that he might just walk past us. It was only three years since Alan Shepard's famous *Freedom 7* flight.

"But we were holding the ace of spades in Dr. Matthew Radnofsky. There just isn't any possibility that he wouldn't stop. He shook hands with Matt, who introduced him to our group. While I was completely in awe of the situation and trying to keep my heart from beating through my shirt, my sister was fascinated that Alan Shepard was wearing shoes and socks like any 'normal' person. He was carrying a briefcase and looked like a fit and trim businessman. I was surprised at the length—and the normalcy—of the very pleasant conversation and his inclusion of all of us.

"We finally said our goodbyes and walked through the doors of the Spacecraft Center. Matt asked us to wait outside one of the inner offices, which he entered. He came out holding a blue flight suit and presented it to me, saying that this was John Glenn's and that he wore it after his *Friendship 7* flight, which had happened in February 1962.

"Through all this, the one thing that stood out in my mind was Alan Shepard's personality. He smiled, laughed and listened. Just perfect, I thought, and I wondered if he would go to the moon!"

Bill Green works with the National Explosives Canine Protection Team of the Transportation Safety Administration/Dept. of Homeland Security at SAIC, and is devoted to community service. He serves as vice-chair of Henderson House, a domestic abuse and sexual assault organization. He is also a board member of the Oregon Juvenile Justice System Children Citizens' Review

Board, making recommendations to the courts concerning children in the foster care system; a volunteer for the Yamhil County Gospel Rescue Mission, providing Domestic and Sexual Violence Outreach Services; and a volunteer for the Evergreen Aviation and Space Museum. Bill's work at the museum includes aircraft restoration and maintenance and educating the volunteers who maintain historic aircraft and prepare new arrivals for display. He also serves as an appointed volunteer advising on airport operational issues.

Horton, Eugene. *Losing Them*. Lulu.com, 2005. A self-described "rambling memoir," in essence a fascinating scrapbook of press releases, opinion pieces, diary entries, stories and reminiscences from NASA's beloved pressman Gene Horton (1930–2021), an Air Force pilot, professional journalist, and longtime public relations officer for NASA, in which he expresses his strong personal opinions and politics. He carefully details protecting Louise and Alan Shepard's privacy after the Mercury astronaut's historic flight so the couple had privacy before their White House meeting with President and Mrs. Kennedy. Gene Horton was a kind friend to countless NASA personnel and their families, including the Radnofsky family.

Johnston, Richard S., James V. Correale, and Matthew I. Radnofsky, "Space Suit Development Status." NASA TN D-3291. Houston, TX: Johnson Space Center, 1966. Written in 1965, this paper discusses advances and plans for space suits, looking to the future for long-term lunar exploration and anticipating extended interplanetary travel, including spacecraft design for sealed-off "emergency donning areas," advocating a "design and operational approach" to make space travel "safe, comfortable and more acceptable to the operating crews."

Jones, H. W. "NASA Should Not Use the Traditional One- or Two-Fault Tolerance Rules to Design for Reliability." 2023 Annual Reliability and Maintainability Symposium (RAMS), Orlando, FL, USA, 2023, pp. 1-6, doi: 10.1109/RAMS51473.2023.10088227. https://ieeexplore.ieee.org/document/10088227.

From the IEEE scientific, safety and engineering standpoint, NASA in the Apollo era spoke to triple redundancy, known also as "two-fault tolerant": "a two-fault tolerant system is one that can operate satisfactorily after experiencing two failures." Published in 2023 Annual Reliability and Maintainability Symposium (RAMS), January 23–26, 2023, Orlando, FL; date added to IEEE Xplore: April 5, 2023. DOI: 10.1109/RAMS51473.2023.10088227.

Kranz, Gene. *Failure Is Not an Option: Mission Control from Mercury to Apollo 13 and Beyond.* New York: Simon & Schuster, 2000. Kranz praises Shepard both for his "marvelously calm" response to the intense moments of Apollo 14 and for his calm and pointed gratitude to Houston for the "nice job done there" with key, successful problem solving. Kranz also made clear that the golfing experiment "did not appear on anyone's manifest, but Shepard had cleared it with Deke Slayton" (349–50).

Lindbergh, Charles A., and Fitzhugh Green. *We.* SeaWolf Illustrated Press edition. New York: G.P. Putnam's Sons, 1927. Asked by Lindbergh to document the reception the homecoming hero received after his transatlantic crossing, writer Fitzhugh Green reflected on a simple ceremony with a brief, extemporaneous speech that Lindbergh gave to the president and multitudes gathered by conveying the "affection of the people" of France and Europe to the people of the United States. The audience, stunned silent for a curiously long time by the brevity of the message, then erupted into clapping, weeping and sobbing, "beginning to realize that something was happening far greater than just the celebration of a mechanical triumph over the ocean separating Europe from America" (162). This handsome reference to community seems to capture the same response the world felt for the early astronaut explorers.

Lovell, Jim, and Jeffrey Kluger. *Lost Moon: The Perilous Voyage of Apollo 13.* 1st ed. Boston: Houghton Mifflin, 1994.

McCall, Robert, and Isaac Asimov. *Our World in Space.* 1st ed. Greenwich, CT: New York Graphic Society, 1974. Magnificent art by NASA personnel's favorite space artist of the Apollo era, painter of history of lunar exploration and the future of space exploration/colonization, vividly explained by Isaac Asimov, a serious scientist as well as famed writer of science fiction.

McKnight, Col. William. Press clip and photo. "Col. McKnight Receives Silver Star for Gallantry in Action—21 Years Ago," *Ramey Tropicair* (Ramey Air Force Base newspaper), November 5, 1965. This is an account of the 1965 discovery of the 1945 commendation letter and description of the gallantry in action by Colonel William McKnight following heavy flak injuring the navigator, Lt. Matt Radnofsky, on November 21, 1944, including young Lt. McKnight's removal of his own parachute and putting it on the injured navigator. The photo shows presentation of the Silver Star—4th highest award by the United States—from the Bombardment Wing Commander to Lt. Col. McKnight. Box 7, March 21, 1966, authors' collection of Radnofsky Papers.

Mindell, David A. *Digital Apollo: Human and Machine in Spaceflight.* Cambridge, MA: MIT Press, 2008. Mindell offers a fascinating, intricately researched view of engineering and human factors in every Apollo lunar pilot landing, as explained by a supremely qualified historian, who emphasizes Shepard's devotion to flight simulation and constant practice. There's something for everyone, as he begins with a philosophical approach to aviation history, citing Antoine de Saint-Exupéry:

> The machine, which at first blush seems a means of isolating man from the great problems of nature, actually plunges him more deeply into them. As for the peasant, so for the pilot, dawn and twilight become events of consequence.

NASA Artemis Project.
https://www.nasa.gov/artemisprogram

"NASA to Dedicate, Rename Rocket Park for Former Director George Abbey." Media Advisory M221-005, December 26, 2021.

https://www.nasa.gov/press-release/nasa-to-dedicate-rename-rocket-park-for-former-director-george-abbey

NASA History. "50 Years Ago: Apollo 14 Crew Leaves Quarantine." NASA release describing intense scheduling planned for travel to speak to the U.S. Congress, meet with the president, appear at celebrations, and travel to the Paris Air show and a meeting with Russian cosmonauts. Shepard looks handsomely groomed in the photo.

https://www.nasa.gov/feature/50-years-ago-apollo-14-crew-leaves-quarantine

NASA. The Longhorn Project. A one-of-a-kind hands-on educational program at Johnson Space Center, The Longhorn Project is a 501(c)(3) charitable nonprofit organization created as a joint venture by NASA's Johnson Space Center, the Houston Livestock Show and Rodeo, Clear Creek Independent School District, and the Texas Longhorn Breeders Association of America. The late George Abbey championed the effort.

www.thelonghornproject.com

Pappas, Terry. *SR-71: The Blackbird, Q&A.* 1st ed. Houston, TX: ADI Publications, 2012. An accomplished writer/Blackbird pilot, Pappas explains what it's like to fly this legendary spy plane, and the great demands that the SR-71 makes on piloting proficiency. In a moving set of questions and answers, Pappas answers why certain people love this kind of work; how it feels to encounter serious physiological problems in the world's fastest aircraft; the greatest challenges; and the importance of practice and the various methods of thinking through practice. He provides great adventures in intelligence gathering. Air Force pilot Pappas was both a Blackbird pilot and a NASA test pilot. Like so many aviators, he's also devoted to golf, with thoughtful insights on the appeal of golf to aviators. Just released in 2023 is a handsome second edition highlighting extraordinary photos from the collections of Lockheed

Martin, the U.S. Air Force, and the author's own collection. Major General Pat Halloran, USAF, ret., who flew in both initial U-2 and SR-71 cadres, opens the book as he praises Pappas as "one of our very best pilots."

———. Personal correspondence to Barbara Radnofsky, December 28, 2023, via email: Pappas remarks to distinguished aviators, honoring Col. Dick Cole (present at the event at age 103), co-pilot of Lt. Col. Jimmy Doolittle in the historic Doolittle Raid:

"I made some remarks on behalf of Col. Dick Cole at the Saturday evening dinner which is held annually for the Wings Over Houston airshow Autograph Tent attendees and their family and friends. This was the summer of 2018, just several months before he passed away at 103. He was in the audience with his daughter, who accompanied him annually. The audience included many famous aviators and aviation support personnel from our military and civilian arenas. George Abbey, former JSC director, was there, along with astronaut Bonnie Dunbar, triple ace fighter pilot Bud Anderson, Marine fighter ace legend Dean Caswell, and many highly distinguished aviators.

"I made my remarks in honor of then 1st Lt. Dick Cole, who everyone knew as then Lt. Col. Jimmy Doolittle's copilot on the first B-25 bomber to launch off the aircraft carrier USS *Hornet*, in April 1942, a mere four months after the attack on Pearl Harbor. No one thought this mission was possible, least of all the Japanese. These bombers had just enough fuel to reach their targets and then crash into the ocean near China. Many of the aircrew were captured and held for months until repatriated. Following his return to US Armed Service, Dick flew more extremely dangerous missions flying the C-47 (DC-3) at night across the Himalayas to deliver needed supplies to Chinese military in support of the Allied war effort. Navigation in these mountains was most perilous for these aircrew.

"Later, Dick volunteered to be among the first pilots to serve as an Air Commando, working behind enemy lines in our war efforts in Burma.

"I saw Dick's eyes light up that night when I pointed out that of the 12 million service members we had in WWII, Dick Cole was the only one to do all three of these dangerous missions."

Pearlman, Robert. Interview by Barbara Radnofsky and Jonathan Richards, Rocket Park, Johnson Space Center, Houston, TX, April 11, 2022. Minor revisions by Pearlman, February 12, 2023. Collections of authors and documentarian.

Radnofsky, Caroline. "Putting Man on the Moon." October 15, 2015. Journalist Caroline Radnofsky explores the legacy of her grandfather, the man behind the iconic Apollo 11 space suits. https://www.aljazeera.com/program/al-jazeera-correspon dent/2015/10/15/putting-man-on-the-moon

Radnofsky, Matthew. Interview by Paul Stallings, Offices of Vinson & Elkins, LLP, Houston, TX, April 9, 1991. Collections of authors and documentarian.

———. "The History of Space Suits and Associated Equipment." Houston, TX, November 10, 1983. Authors' collection of Radnofsky Papers. This brief history advocates international cooperative data-sharing on space suit materials testing and sources, with comments on the famous blue flight suit that was "not carried aboard the Mercury Capsule, but was included in a special clothing kit that was carried aboard the recovery ships." The paper notes that a Mercury astronaut had no room to remove his space suit, which was "meant only to be a back-up to the cabin pressurized system, a 'get-me-down' device. The fact is, no astronaut ever inflated a Mercury suit in space!"

———. NASA publications, interviews, patents. https://ntrs.nasa.gov/search?q=radnofsky

———. NASA Conference on Materials for Improved Fire Safety, NASA Technical Reports Server (NTRS) https://ntrs.nasa.gov/citations/19700024901

———. Radnofsky Apollo Suits Oral History Transcript. Hous-

ton: Box 7, March 21, 1966, Authors' collection of Radnofsky
Papers.

https://uhcl-ir.tdl.org/bitstream/handle/10657.1/1086/jsc-
center-space-suits.pdf?sequence=1&isAllowed=y

Richards, Jonathan. Website of documentarian. The website
contains links to a wide variety of the documentarian's films,
including his documentary covering Matt Radnofsky's work at
NASA.

www.jonathanrichards.tv

———. *The Barber, the Astronaut and the Golf Ball.* Vimeo link
to the documentary film of the same title as this book, on website of
documentarian. Jonathan Richards owns all rights to the film.

https://vimeo.com/jonathanrichards/bagb

Roach, Joe. Interview, NASA Oral History project, January 24,
2000. Roach noted that listening to the radio detailing Alan Shepard's Freedom 7 flight motivated him to pursue a career in the
space program (6). He tells a story on himself: he correctly anticipated the first question Kris Kraft would ask of Roach after they
received the call of the Apollo 13 danger. The brilliant Mr. Roach
had anticipated and planned for the lunar lander to be used as a
lifeboat to allow the crew to return to Earth. Kraft: "I guess you've
got your lifeboat procedures." Roach: [Laughter] "Roger. We did."

https://historycollection.jsc.nasa.gov/JSCHistoryPortal/
history/oral_histories/RoachJW/JWR_1-24-2000.pdf

Schefter, James. *The Race: The Uncensored Story of How
America Beat Russia to the Moon.* New York: Doubleday, 1999.
Schefter covered NASA in the 1960s for the *Houston Chronicle*,
then for *Time-Life*, giving him special access thanks to *Life*'s exclusive contract with the Original Seven astronauts. He provides
detail from 1961 on attributes of the Manned Spacecraft Center
location as meeting all the government's criteria, as well as of the
politics and deals behind the site selection (149–50). Schefter
used NASA materials and personal papers of Russian and U.S.
participants, as well as some of his experiences, to provide

"behind-the-scenes" accounts of the astronauts and the space race. His writing soars in describing Alan Shepard as an astronaut in the Original Seven. Shepard emerges as "probably the smartest" and "the one who always saw the big picture," "extraordinarily observant" and a "major contributor" in lengthy meetings or in "the adrenaline rush of postflight debriefings" (71). After his historic Freedom 7 flight, Shepard "was brilliant" during debriefing sessions. "He remembered everything," and he "gave detailed and expansive answers. . . . He was as professional as an engineer and a test pilot could be." Perhaps most moving is the journalist's account of a 1967 banquet honoring Shepard, on the sixth anniversary of his famous suborbital flight, organized to "do something positive" after the tragedy of the Apollo 1 fire, with all proceeds donated to the Ed White Memorial Fund. Shepard agreed to be honored and even sang in a Vaudeville routine before speaking movingly to comfort and inspire colleagues mourning the loss of the Apollo 1 astronauts Grissom, White, and Chafee (251–53).

Shepard, Alan B. *Training by Simulation.* Edwin A. Link Lecture Series. Washington D.C.: Smithsonian Institution, 1965. Shepard delivered the first annual Edwin A. Link Lecture on February 19, 1964, with the famous pioneer in training pilots in the original "Link Trainer." He lauded the more flexible and accurate computers that opened a new field. He emphasized the value of the key element of training for manual control in Mercury, where automation became key. As it turned out, manual control played a more important role in accomplishing a successful program than the automatic systems (5–6). As Shepard praised the original Link trainers teaching instrument flying, so he praised the advanced simulators "helping us develop, day by day, a safer and more carefully developed planned program for accomplishing our goals in space" (12).

———. Interview by *The Academy of Achievement*, February 1, 1981, www.achievement.org. A wide-ranging interview reflecting on Shepard's personal as well as professional life, family, background, and famous role of moon-golfer.

https://m.youtube.com/watch?v=G3tFb543lsg

————. Interview by Roy Neal, Johnson Space Center Oral History Project, February 20, 1998. Roy Neal engaged Alan Shepard in a charming, reflective, thoughtful and—in retrospect—poignant, interview. The astronaut said landing on the moon (the "moon deal") was a "piece of cake compared to landing a plane on a carrier." He emphasized the importance of practice as a key in spaceflight as well as flying commercial aircraft; creating a "sense of confidence," he said, is "very important to a pilot." He said he practiced the golf swing on earth to ensure there were no safety issues. As with the *Academy* interview, he said he brought two balls to the surface (although, as memorabilia expert Pearlman observed, his first wording on the moon implied but one ball . . . until he whiffed the first ball and produced another.) When Neal asked about the value of John Glenn flying at an advanced age, Shepard said he thought it a good idea, referencing information gained on what would happen to older bones. Asked if he wanted to fly again, he smiled poignantly, responding—with a catch in his voice—that he had a health condition which would prevent him doing so. Shepard passed away that same year, in July 1998. His beloved wife Louise passed away five weeks later.

https://youtu.be/Oc7HVfjLsiU

Shepard, Alan B., and Deke Slayton. *Moon Shot: The Inside Story of America's Race to the Moon.* 1st ed. Atlanta, GA: Turner Publications, 1994. Revealing autobiographies of both men giving details of their lives and missions reflecting critical times.

Smylie, Ed. Interview, NASA Oral History Project. https://historycollection.jsc.nasa.gov/JSCHistoryPortal/ history/oral_histories/SmylieRE/RES_4-17-99.pdf

Strong, Russell A. *First Over Germany: A History of the 306th Bombardment Group.* Winston Salem, NC: Hunter Publishing Company, 1982. This group was one of the oldest flying missions over Nazi-occupied Europe in 1942–45, and Strong details the mission leading to capture of 1st Lt. Matthew Radnofsky and McKnight. Flak hit under the nose, wounding navigator Radnofsky, another blast injured the co-pilot, and the explosions set the number two engine on fire and severed the throttle linkage to the number one engine. The captain had to order the crew to bail out. The pilot landed the crippled plane in Firsteman, Germany. The captured POWs missed Thanksgiving observance at their home base in Bedford, where Lt. General James Doolittle read the presidential Thanksgiving Proclamation at St. Paul's Church (289–90).

Supkis, Daniel E. *Description and Applications of Fluorel L-3203-6.* NASA Technical Reports Service, Technical Memorandum 19700009416, MSC 01275, January 1, 1969. The NASA files contain great photos of Fluorel® coating on circuit breakers and printed circuit boards, which appear to have been perfectly duplicated—or used, perhaps—in the Tom Hanks *Apollo 13* movie.

https://ntrs.nasa.gov/search?q=supkis

Thompson, Neal. *Light This Candle: The Life and Times of Alan Shepard, America's First Spaceman.* 1st ed. New York: Crown Publishers, 2004. Upon Alan Shepard's death in 1998, finding no biography of America's first man in space except a 1962 young adult book, *Baltimore Sun* journalist Neal began an intensively researched biography, notable for its explanation of Shepard's great generosity toward Deke and Marge Slayton—and to the world—in his commitment to the dual memoir by Slayton and Shepard, *Moon Shot.* Shepard had zealously guarded his privacy in his personal life. But when he learned that Slayton had been diagnosed with a brain tumor and that his prospects did not look good, Shepard agreed: "If this will help Deke, I'll do it." Thompson explained further in a fine chapter devoted to Shepard's quiet generosity, including vigorous promotion of the dual memoir *Moon Shot* and its revelations of the wonderful friendship between the two great aviators. Slayton's widow, Bobbie, went on record to thank Shepard, acknowledging his kindness and great friendship: "He did that for Deke and me" (389). Thompson's varied sources include more than 400 pages of an FBI "extensive background investigation of Shepard in 1971, at a time when he was under consideration for a presidential appointment," and a 1967 Civil Service Commission background investigation. Although Thompson boldly claims that Shepard's "moon balls were, in fact, driving range balls made by Spalding" (280), he cites for this proposition the claim of the son of a River Oaks Country Club pro, stating that his dad supplied particularly durable balls, which read "Property of Jack Harden." See Jaime Diaz, "Shooting for the Moon" (earlier in these sources) for various opinions on how golf balls would endure conditions on the moon surface, and see also the Shepard interview by *The Academy of Achievement* (also earlier in these sources) for Shepard's expectations. Shepard never revealed the brand of balls he took to the moon, but he did predict that his successors would find the balls and play with them.

Villagomez, Carlos. Interview by Barbara Radnofsky and Jonathan Richards, Carlos Beer Garden and Barbershop, Webster, TX, April 14, 2022. Collections of authors and documentarian.

———. Press clip. Lee Holley, "Shepard's Hairdresser Knows His Astro-follicles," *News Citizen*, February 10, 1971. The *News Citizen* and *Dallas Morning News* both interviewed Carlos in depth, learning of the post-quarantine haircut well in advance and getting extraordinary details. Collection of authors.

———. Press clips. "Shepard's Cut: 'Shag Look' Very Popular," and "Staff Special in the News," *Dallas Morning News*, February 10, 1971, p. 16. A day after splashdown and safe arrival home of the Apollo 14 astronauts, and a month after pre-launch isolation had begun on January 11, Shepard "spoke from the heart" to flight director Gerald Griffin: "I sure do want to thank you for that superb job you did for us. It's a hell of a thrill for us to work with you." The paper said Carlos Villagomez had watched the splashdown on TV, and that Shepard would be "sitting in Villagomez' chair less than three weeks from now when lunar quarantine is lifted." Additional background told readers of the proud display of a "Go Navy" sign in the seat used by Shepard. "Visiting Shepard's home last month, before the Apollo 14 crew went into pre-flight quarantine, Villagomez cut the astronaut's dark brown hair short enough to last 45 days. 'If it's not cut and combed right, it looks terrible,' Villagomez said. 'He has a stubborn type of hair, hard to do anything with.'" Collection of authors.

———. Press clip. "Shepard's Hunt for a Hairdo," *San Francisco Chronicle,* February 5, 1971, p. 9. Friday, February 5, was the date of the Apollo 14 lunar landing. *Chronicle* coverage explained that Shepard had paid Carlos "$50 a year ago" to find him a new style to replace the crew cut he had worn since before his first space flight ten years before: "He'd been wanting to do it for a long time but didn't want to go through that bad state. We tried three, four styles. We left it flat on top but it stuck straight out. It's stubborn hair. We let it grow a lot longer but that was no good because he couldn't control it. He's always wearing helmets. We put a part in it but that made him look sort of strange. . . . It made him look more youthful. The man, close up, doesn't look 30." The *Chronicle* quoted Carlos on his client's hair: "Shepard is most conscious of his hair. He said the astronaut's secretary made appointments for Shepard and he has flown back from Cape Canaveral for a $6 haircut." The paper provided much detail—"The new style is razor cut, 1½ inches long all over the head. It doesn't need combing"—and concluded that the astronaut had "a very full head of hair for a 47- year-old man. A lot of people think he dyed it, but he didn't." Collection of authors.

———. Press clip. "Hairdresser Tells ALL on Shepard," *Los Angeles Times*, February 5, 1971, p. 1R. This gave some of the same United Press information as appeared in the *Washington Post* on February 6, 1971, p. E4, quoting Carlos complimenting Shepard for being conscious of his hair. The *Times* and *Post* ran variations quoting Carlos on Shepard: "He's a cat, man. He drives real fast cars. He dresses sharp. He's an up to date fellow." The *Post* explained the trial periods for various Shepard hairstyles, quoting

Carlos describing Shepard as "the Joe Namath of astronauts." Collection of authors.

Wolfe, Tom. *The Right Stuff.* Illustrated ed. New York: Black Dog and Leventhal/Workman Publishing, 2004 [1979]. "The goal in Project Mercury, as in every important new flight project, was to be the pilot assigned to make the first flight. In flight test that meant your superiors looked upon you as *the* man who had the right stuff to challenge the unknowns." (82, emphasis in original).

ABOUT THE AUTHORS

Barbara Radnofsky and Ed Supkis grew up in the 1960s in the shadow of NASA's Manned Spacecraft Center and married in 1982. They have three children and five grandchildren. The couple —with many other community members — are co-owners of Brazos Bookstore, an independent bookseller.

As children of NASA scientists, Barbara, and Ed had front-row seats to the development of the space program and the community built around it on rural cow pastures near Webster, Texas.

———

BARBARA RADNOFSKY is a mediator, teacher and lawyer. She was named the Outstanding Young Lawyer of Texas in 1988. Practicing on both sides of the docket, she's been listed for more than thirty years in "Best Lawyers in America" in multiple areas. She's the author of *A Citizen's Guide to Impeachment*, a nonpartisan explanation of U.S. constitutional impeachment history and practice.

Barbara co-founded the Houston chapter of the National Association of Urban Debate

Leagues. She's served on many other charitable boards and as a volunteer peer mediation teacher in public and private schools.

———

ED SUPKIS MD is a board-certified anesthesiologist specializing in cardiac anesthesia, and he worked with Dr. Michael DeBakey and his associates for over a decade. He then practiced anesthesia at MD Anderson Cancer Center, where he served as Director of Quality Assurance for the Division of Anesthesiology and as Medical Director of Respiratory Care for the Division of Surgery and Anesthesiology. He later served as Medical Director for the Houston Methodist Hospital Anesthesia Preoperative Assessment Clinic where he was awarded the Hospital's Patient Safety Award.

Throughout his anesthesia career, Ed worked on numerous national and international committees, devoted to quality improvement, standards, testing, and patient safety. Among his many inventions, he is best known for the Supkis Catheter, a specialized, patented device used for safe exchange of breathing tubes in patients with difficult airways.

He currently serves as a Senior Aviation Medical Examiner for the Federal Aviation Administration. Ed is also an instrument-rated, multi engine, private pilot.

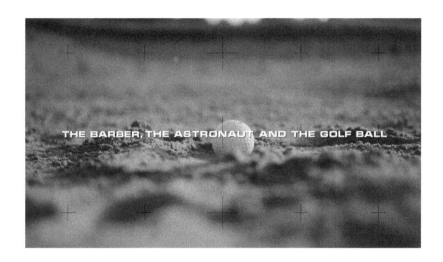

THE BARBER, THE ASTRONAUT, AND THE GOLF BALL

Watch the documentary *The Barber, The Astronaut, and The Golf Ball* directed by Jonathan Richards

Looking for your next book?
We publish the stories you've been waiting to read!

A Member of the Texas Book Consortium

Check out our other titles, including audio books, at
StoneyCreekPublishing.com.

For author book signings, speaking engagements, or other events,
please contact us at info@stoneycreekpublishing.com